MEMOIRS
of the
American Mathematical Society

Volume 230 • Number 1079 (first of 5 numbers) • July 2014

Formality of the Little N-disks Operad

Pascal Lambrechts
Ismar Volić

ISSN 0065-9266 (print) ISSN 1947-6221 (online)

American Mathematical Society
Providence, Rhode Island

Library of Congress Cataloging-in-Publication Data

Lambrechts, Pascal, 1964-
Formality of the little N-disks operad / Pascal Lambrechts, Ismar Volić.
pages cm. – (Memoirs of the American Mathematical Society, ISSN 0065-9266 ; volume 230, number 1079)
Includes bibliographical references and index.
ISBN 978-0-8218-9212-1 (alk. paper)
1. Homotopy theory. 2. Operads. 3. Loop spaces. I. Volic, Ismar, 1973- II. Title.

QA612.7.L36 2014
514′.24–dc23 2014008206

DOI: http://dx.doi.org/10.1090/memo/1079

Memoirs of the American Mathematical Society

This journal is devoted entirely to research in pure and applied mathematics.

Subscription information. Beginning with the January 2010 issue, *Memoirs* is accessible from www.ams.org/journals. The 2014 subscription begins with volume 227 and consists of six mailings, each containing one or more numbers. Subscription prices are as follows: for paper delivery, US$827 list, US$661.60 institutional member; for electronic delivery, US$728 list, US$582.40 institutional member. Upon request, subscribers to paper delivery of this journal are also entitled to receive electronic delivery. If ordering the paper version, add US$10 for delivery within the United States; US$69 for outside the United States. Subscription renewals are subject to late fees. See www.ams.org/help-faq for more journal subscription information. Each number may be ordered separately; *please specify number* when ordering an individual number.

Back number information. For back issues see www.ams.org/bookstore.

Subscriptions and orders should be addressed to the American Mathematical Society, P. O. Box 845904, Boston, MA 02284-5904 USA. *All orders must be accompanied by payment.* Other correspondence should be addressed to 201 Charles Street, Providence, RI 02904-2294 USA.

Memoirs of the American Mathematical Society (ISSN 0065-9266 (print); 1947-6221 (online)) is published bimonthly (each volume consisting usually of more than one number) by the American Mathematical Society at 201 Charles Street, Providence, RI 02904-2294 USA. Periodicals postage paid at Providence, RI. Postmaster: Send address changes to Memoirs, American Mathematical Society, 201 Charles Street, Providence, RI 02904-2294 USA.

This publication is indexed in *Mathematical Reviews*®, *Zentralblatt MATH*, *Science Citation Index*®, *Science Citation Index*™*-Expanded*, *ISI Alerting Services*℠, *SciSearch*®, *Research Alert*®, *CompuMath Citation Index*®, *Current Contents*®*/Physical, Chemical & Earth Sciences*. This publication is archived in *Portico* and *CLOCKSS*.
Printed in the United States of America.

∞ The paper used in this book is acid-free and falls within the guidelines established to ensure permanence and durability.
Visit the AMS home page at http://www.ams.org/

10 9 8 7 6 5 4 3 2 1 19 18 17 16 15 14

Contents

Abstract

The little N-disks operad, $\mathcal{B}$, along with its variants, is an important tool in homotopy theory. It is defined in terms of configurations of disjoint N-dimensional disks inside the standard unit disk in $\mathbb{R}^N$ and it was initially conceived for detecting and understanding N-fold loop spaces. Its many uses now stretch across a variety of disciplines including topology, algebra, and mathematical physics.

In this paper, we develop the details of Kontsevich's proof of the formality of little N-disks operad over the field of real numbers. More precisely, one can consider the singular chains $\mathrm{C}_*(\mathcal{B};\mathbb{R})$ on $\mathcal{B}$ as well as the singular homology $\mathrm{H}_*(\mathcal{B};\mathbb{R})$ of $\mathcal{B}$. These two objects are operads in the category of chain complexes. The formality then states that there is a zig-zag of quasi-isomorphisms connecting these two operads. The formality also in some sense holds in the category of commutative differential graded algebras. We additionally prove a relative version of the formality for the inclusion of the little m-disks operad in the little N-disks operad when $N \geq 2m+1$.

The formality of the little N-disks operad has already had many important applications. For example, it was used in a solution of the Deligne Conjecture, in Tamarkin's proof of Kontsevich's deformation quantization conjecture, and in the work of Arone, Lambrechts, Turchin, and Volić on determining the rational homotopy type of spaces of smooth embeddings of a manifold in a large euclidean space, such as the space of knots in $\mathbb{R}^N$, $N \geq 4$.

Received by the editor January 25, 2011, and, in revised form, July 12, 2012.

Article electronically published on November 14, 2013.

DOI: http://dx.doi.org/10.1090/memo/1079

2010 *Mathematics Subject Classification.* Primary 55P62; Secondary 18D50.

Key words and phrases. Operad formality, little cubes operad, Fulton-MacPherson operad, trees, configuration space integrals.

The first author is Maître de Recherches au F.R.S-FNRS..

The second author was supported in part by the National Science Foundation grants DMS 0504390 and DMS 1205786.

Affiliations at time of publication: Pascal Lambrechts, Université catholique de Louvain, IRMP 2 Chemin du Cyclotron, B-1348 Louvain-la-Neuve, Belgium, email: pascal.lambrechts@uclouvain.be; and Ismar Volić, Department of Mathematics, Wellesley College, Wellesley, Massachusetts 02482; email: ivolic@wellesley.edu.

Acknowledgments

Our deepest gratitude goes to Greg Arone for his encouragement, support, and patience. We also thank Victor Turchin for his encouragement and for explaining the proof of Theorem 8.1 to us. We also thank Nathalie Wahl for pointing out some errors and some weaknesses in exposition in an earlier version of this paper. We are grateful to Paolo Salvatore for pointing out to us the reference [**21**, Lemma 6.4]. Parts of this paper were written while the first author was visiting the University of Virginia, Wellesley College, and the Center for Deformation and Symmetry at University of Copenhagen and while the second author was visiting University of Louvain, Massachusetts Institute of Technology, and the University of Virginia. We would like to thank these institutions for their hospitality and support. Lastly, we wish to thank the referee for a thorough reading of the paper and for helpful suggestions and comments.

CHAPTER 1

Introduction

In this paper we give a detailed proof of Kontsevich's theorem on the formality of the little N-disks operad. The theorem, whose proof was sketched in [**20**, Theorem 2], asserts that the singular chains on the little N-disks operads is weakly equivalent to its homology in the category of operads of chain complexes. We also improve that result in three directions:

(1) Formality is in the category of CDGA (commutative differential graded algebras) which, following Sullivan and Quillen, models rational homotopy theory;
(2) For us, the little disks operad has an operation in arity 0 while Kontsevich discards that nullary operation;
(3) We establish a *relative* formality result, namely formality of the inclusion of the little m-disks operad into the little N-disks operad for $N \geq 2m+1$.

Our motivation for proving these results comes from applications to the study of the rational homology of the space $\mathrm{Emb}(M, \mathbb{R}^N)$ of smooth embeddings of a compact manifold M into $\mathbb{R}^N$. Goodwillie-Weiss manifold calculus [**16**,**32**] approximates this embedding space by homotopical constructions based on a category $\mathcal{O}_\infty$ of open subsets of M diffeomorphic to finitely many open balls with inclusions as morphisms. This category is closely related to the little balls operad. On the other hand, formality theorems can often lead to collapse results for spectral sequences. Combining manifold calculus with formality, the authors, along with Greg Arone, were thus able to prove in [**3**] the collapse of a spectral sequence computing $\mathrm{H}_*(\overline{\mathrm{Emb}}(M, \mathbb{R}^N); \mathbb{Q})$, where $\overline{\mathrm{Emb}}(M, \mathbb{R}^N)$ is a slight variation of $\mathrm{Emb}(M, \mathbb{R}^N)$. A special case of this approach also led the authors, jointly with Victor Turchin, to the proof in [**23**] of the collapse of the Vassiliev spectral sequence computing the rational homology of the space of long knots in $\mathbb{R}^N$ for $N \geq 4$.

To explain the formality results that we prove here, fix an integer $N \geq 1$ and recall the classical little N-disks operad $\mathcal{B}_N = \{\mathcal{B}_N(n)\}_{n \geq 0}$, where $\mathcal{B}_N(n)$ is the space of configurations of n closed N-disks with disjoint interiors contained in the unit disk of $\mathbb{R}^N$ [**4**]. The integer N will usually be understood so we will just denote this operad by $\mathcal{B}$ and often simply say "little balls operad". This operad is homotopy equivalent to many other operads, such as the little N-cubes operad, or the Fulton-MacPherson operad $\mathrm{C}[\bullet] = \{\mathrm{C}[n]\}_{n \geq 0}$ of compactified configurations of points in $\mathbb{R}^N$. The latter will be important in our proofs and we will say more about it in Chapter 5.

Fix a unital commutative ring $\mathbb{K}$. The functor

$$\mathrm{S}_*(-; \mathbb{K}) \colon \mathrm{Top} \longrightarrow \mathrm{Ch}_{\mathbb{K}}$$

of singular chains with coefficients in $\mathbb{K}$ is symmetric monoidal. Therefore $\mathrm{S}_*(\mathcal{B}; \mathbb{K})$ is an operad of chain complexes. In addition, its homology $\mathrm{H}_*(\mathcal{B}; \mathbb{K})$ can be viewed as an operad of chain complexes with differential 0. One of the main results that we will prove in detail is

THEOREM 1.1 (Kontsevich [**20**]; Tamarkin for $N = 2$ [**31**]). *The little N-disks operad is stably formal over the real numbers, that is, there exists a chain of weak equivalences of operads of chain complexes*

$$\mathrm{S}_*(\mathcal{B}_N;\mathbb{R}) \xleftarrow{\simeq} \cdots \xrightarrow{\simeq} \mathrm{H}_*(\mathcal{B}_N;\mathbb{R}).$$

The proof of this theorem was sketched in [**20**, Section 3.3] but we felt that it would be useful to develop it in full detail. In this paper, $\mathcal{B}(0)$ is the one-point space, contrary to [**20**] where it is the empty set. This fact makes our proof more delicate, but in the application we have in mind it will be important to have $\mathcal{B}(0) = *$ (operad composition with this corresponds to the operation of forgetting a ball from a configuration of little balls).

Morally, singular chains with coefficients in $\mathbb{Q}$ encode the rational stable homotopy type of spaces or topological operads, and with coefficients in $\mathbb{R}$ we get the "real stable homotopy type". This is why in Theorem 1.1 we talk about *stable* formality. The unstable real (or more correctly, rational) homotopy type of spaces is encoded by commutative differential graded algebras (CDGAs for short), as was discovered by Sullivan using the functor A_{PL} of polynomial forms (see Chapter 3). One then has the important notion of a *CDGA model* for a space X, which by definition is a CDGA weakly equivalent to $A_{PL}(X)$. Any CDGA model (over the field $\mathbb{Q}$) for a simply-connected space with finite Betti numbers contains all the information about its rational homotopy type. We can define an analogous notion of a CDGA model for a topological operad, although the definition is a little bit more intricate (see Definition 3.1). We then have the following unstable version of Theorem 1.1.

THEOREM 1.2. *For $N \neq 2$, a CDGA model over $\mathbb{R}$ of the little N-disks operad is given by its cohomology algebra, that is, it is formal over $\mathbb{R}$ (in the sense of Definition 3.1).*

As explained in Chapter 3, one reason for which our definition of a CDGA model for an operad is not as direct as one might wish is that $A_{PL}(\mathcal{B})$ is not a cooperad. This is because the contravariant functor A_{PL} is not comonoidal. It might be better to consider the coalgebra of singular chains $\mathrm{S}_*(\mathcal{B};\mathbb{R})$, which is indeed an operad of differential coalgebras. However, we do not know how to prove that this operad is weakly equivalent to its homology in the category of differential coalgebras. Moreover, that category is not very suitable for doing real homotopy theory because of the lack of strict cocommutativity.

In Theorem 1.2, we assumed $N \neq 2$. Our proof in the case $N = 2$ fails because some of our CDGAs become $\mathbb{Z}$-graded instead of non-negatively graded as required in rational homotopy theory. We still however obtain some results in the case $N = 2$ and we believe that our proof can be adapted to include that case as well; see Chapter 10.

We now state a relative version of the above theorems. Let $1 \leq m \leq N$ be integers and suppose given a linear isometry

$$\epsilon\colon \mathbb{R}^m \longrightarrow \mathbb{R}^N.$$

Define the map

$$\mathcal{B}_\epsilon(n)\colon \mathcal{B}_m(n) \longrightarrow \mathcal{B}_N(n)$$

that sends a configuration of n m-disks to the configuration of n N-disks where the center of each N-disk is the image under ϵ of the center of the corresponding m-disk and has the same radius. This clearly defines a morphism of operads.

DEFINITION 1.3. A morphism of topological operads

$$\alpha\colon \mathcal{A} \longrightarrow \mathcal{A}'$$

is *stably formal* over $\mathbb{K}$ if there exists a zig-zag of quasi-isomorphisms of operads in $\mathrm{Ch}_{\mathbb{K}}$ connecting the singular chains $\mathrm{S}_*(\alpha;\mathbb{K})$ to its homology $\mathrm{H}_*(\alpha;\mathbb{K})$ as in the following diagram:

$$\begin{array}{ccccccccc}
\mathrm{S}_*(\mathcal{A};\mathbb{K}) & \xleftarrow{\simeq} & \mathcal{C}_1 & \xrightarrow{\simeq} & \cdots & \xleftarrow{\simeq} & \mathcal{C}_k & \xrightarrow{\simeq} & \mathrm{H}_*(\mathcal{A};\mathbb{K}) \\
\downarrow{\scriptstyle \mathrm{S}_*(\alpha)} & & \downarrow & & & & \downarrow & & \downarrow{\scriptstyle \mathrm{H}_*(\alpha)} \\
\mathrm{S}_*(\mathcal{A}';\mathbb{K}) & \xleftarrow{\simeq} & \mathcal{C}'_1 & \xrightarrow{\simeq} & \cdots & \xleftarrow{\simeq} & \mathcal{C}'_k & \xrightarrow{\simeq} & \mathrm{H}_*(\mathcal{A}';\mathbb{K})
\end{array}$$

When $\mathbb{K}$ is a field of characteristic 0, we say that α is *formal* over $\mathbb{K}$ if the morphism of CDGA cooperads $\mathrm{H}^*(\alpha;\mathbb{K})$ is a model for α (see Chapter 3 for the precise definition of a model for CDGA cooperads).

THEOREM 1.4. *Assume that $m \geq 1$ and $N \geq 2m+1$. Then the morphism of operads*

$$\mathcal{B}_\epsilon\colon \mathcal{B}_m \longrightarrow \mathcal{B}_N$$

is stably formal over $\mathbb{R}$. If $m \neq 2$, it is also formal over $\mathbb{R}$.

There is also a notion of *coformality* which is Eckman-Hilton dual to that of (unstable) formality [**24**]. Roughly speaking, coformality of a space X means that its rational homotopy type is determined by its rational homotopy Lie algebra $\pi_*(\Omega X)\otimes\mathbb{Q}$ (instead of its rational cohomology algebra in the case of formality). In some sense, the operad of little N-disks also seems to be coformal, although there is difficulty in making this idea precise because of the lack of a basepoint for the operad. We refer the reader to [**2**] for a discussion of coformality of the little N-disks operad.

All of the above formality results are over the field of real numbers. It would be more convenient to have rational formality because localization over $\mathbb{Q}$ is topologically meaningful, contrary to localization over $\mathbb{R}$. This descent of fields for stable formality of operads is always possible when one considers operads in which the zeroth term (corresponding to 0-ary operations) is empty, as proved in [**17**, Theorem 6.2.1]. In particular, we can consider the operad $\widetilde{\mathcal{B}}$ defined by $\widetilde{\mathcal{B}}(0)=\emptyset$ and $\widetilde{\mathcal{B}}(n)=\mathcal{B}(n)$ for $n\geq 1$. Our formality results for $\mathcal{B}$ are clearly also true for $\widetilde{\mathcal{B}}$; the latter was the operad considered by Kontsevich in [**20**]. Moreover, since this operad has no nullary operations, stable formality for $\widetilde{\mathcal{B}}$ over $\mathbb{R}$ descends to $\mathbb{Q}$.

For our applications to embedding spaces [**3, 23**], however, it is important to take the usual little balls operad, $\mathcal{B}$, which is only formal over $\mathbb{R}$. In those applications, this weaker formality is sufficient essentially because the main results there are about collapse of spectral sequences, and these collapse results do not depend on which field of characteristic 0 is used. The proof of descent of formality in [**17**, Section 6] does not generalize easily to the case with nullary operations because of the lack of minimal models when these degeneracy operations occur.

The formality of the operad $\widetilde{\mathcal{B}}$ implies the formality over $\mathbb{Q}$ of each space $\mathcal{B}(n)$, in the sense that the CDGA $A_{PL}(\mathcal{B}(n))$ is weakly equivalent to its cohomology algebra,

$H^*(\mathcal{B}(n); \mathbb{Q})$. Paolo Salvatore has recently proved using a computer that, for $n = 4$ and $N = 2$, the space $\mathcal{B}_2(4)$ is not formal over the ring $\mathbb{Z}/2$, i.e. its cohomology algebra, $H^*(\mathcal{B}_2(4); \mathbb{Z}/2)$, and its algebra of singular cochains, $S^*(\mathcal{B}_2(4); \mathbb{Z}/2)$, are not quasi-isomorphic. We do not know whether the (non-symmetric) little disks operad is stably formal over some field of positive characteristic.

As a final comment, the Tamarkin's and Kontsevich's proofs of formality for $N = 2$ have been compared in [**26**] where it is proved that the weak equivalences obtained in those two proofs are homotopic.

We end this introduction by explaining the general idea of Kontsevich's proof of formality that we develop in this paper. The main ingredient is a combinatorial CDGA cooperad $\mathcal{D} = \{\mathcal{D}(n)\}_{n\geq 0}$ of *admissible diagrams* and an explicit CDGA map

$$\mathrm{I}\colon \mathcal{D}(n) \longrightarrow \Omega_{PA}(\mathrm{C}[n]) \tag{1.1}$$

which we will call the *Kontsevich configuration space integral*. Here $\mathrm{C}[n]$ are compact manifolds homotopy equivalent to $\mathcal{B}(n)$, and Ω_{PA} is a semi-algebraic analog of the deRham CDGA of differential forms Ω_{DR}. A combinatorial argument will show that the cooperad $\mathcal{D}$ is quasi-isomorphic to the cohomology of the little balls operad. We will also show that I is a quasi-isomorphism and, since I also respects the cooperad structures, the desired result will follow.

Let us elaborate on $\mathcal{D}(n)$ and I a bit further. We will work with the Fulton-MacPherson operad $\mathrm{C}[\bullet] = \{\mathrm{C}[n]\}_{n\geq 0}$ which is homotopy equivalent to the little balls operad. The space $\mathrm{C}[n]$ is a compact manifold with corners obtained by adding a boundary to the open manifold $F_n(\mathbb{R}^N)$, the space of configurations of n points in $\mathbb{R}^N$, that is,

$$F_n(\mathbb{R}^N) := \{(z_1, \dots, z_n) \in (\mathbb{R}^N)^n : z_i \neq z_j \text{ for } i \neq j\}$$

(after normalizing by modding out by translations and positive dilations). Arnold [**1**] computed the cohomology algebra of $F_n(\mathbb{R}^2) = F_n(\mathbb{C})$ and in fact proved that these spaces are formal over $\mathbb{C}$. His argument is as follows:

Consider the complex smooth differential one-forms

$$\omega_{ij} := \frac{d(z_j - z_i)}{z_j - z_i} = d\log(z_j - z_i) \in \Omega^1_{DR}(F_n(\mathbb{C}); \mathbb{C}) \tag{1.2}$$

which are cocycles and can easily be shown to be cohomologically independent for $1 \leq i < j \leq n$. A direct computation shows that these forms satisfy the 3-*term relation*

$$\omega_{ij} \wedge \omega_{jk} + \omega_{jk} \wedge \omega_{ki} + \omega_{ki} \wedge \omega_{ij} = 0. \tag{1.3}$$

It is convenient to represent this relation by the diagram pictured in Figure 1.1.

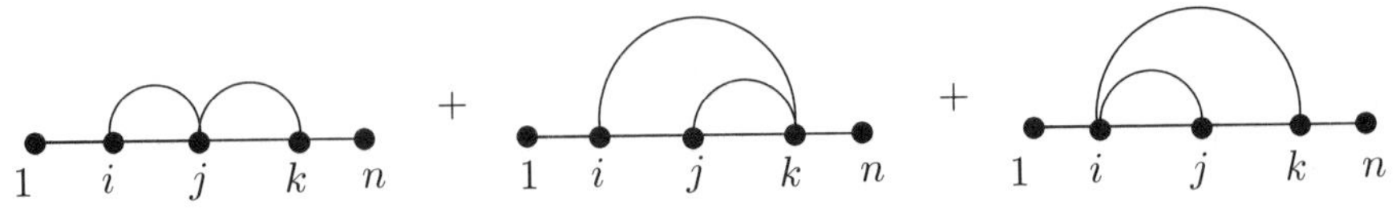

FIGURE 1.1. Diagrammatic description of the 3-term relation.

In this figure, the vertices on the line correspond to the labels of the points $z_1, \dots, z_n$ of a configuration and each edge (u, v) between two vertices represents a differential form ω_{uv}.

The subalgebra of $\Omega_{DR}(F_n(\mathbb{C}); \mathbb{C})$ generated by the ω_{ij} is

$$\frac{\wedge(\omega_{ij} : 1 \le i < j \le n)}{(\omega_{ij} \wedge \omega_{jk} + \omega_{jk} \wedge \omega_{ki} + \omega_{ki} \wedge \omega_{ij})}.$$

This algebra has a trivial differential and it maps to the cohomology algebra $\mathrm{H}^*(F_n(\mathbb{C}); \mathbb{C})$. A Serre spectral sequence argument shows that this map is actually an isomorphism. In other words, the cohomology embeds in the deRham algebra of forms, and hence $F_n(\mathbb{C})$ is formal.

Arnold's argument for $N = 2$ can be generalized to all N as follows. Consider the differential forms $\omega_{ij} = \theta_{ij}^*(\mathrm{vol})$ where

$$\begin{aligned} \theta_{ij} \colon \quad F_n(\mathbb{R}^N) &\longrightarrow S^{N-1} \\ (z_1, \dots, z_n) &\longmapsto \frac{z_j - z_i}{\|z_j - z_i\|}, \end{aligned}$$

and $\mathrm{vol} \in \Omega_{DR}^{N-1}(S^{N-1})$ is the symmetric volume form on the sphere S^{N-1} that integrates to 1. For $N = 2$, these are analogous to (1.2). It is well known by work of F. Cohen that these forms generate the cohomology algebra of $F_n(\mathbb{R}^N)$ and that the 3-term relation holds in cohomology. However, the relation is not always true at the level of forms. One only knows that, for each i, j, and k, there exists some differential form β such that

$$d\beta = \omega_{ij} \wedge \omega_{jk} + \omega_{jk} \wedge \omega_{ki} + \omega_{ki} \wedge \omega_{ij}. \tag{1.4}$$

The key idea now is to describe an algorithm which constructs *in a natural way* such a cobounding form β. To explain this, suppose that $n = 3$ and $(i, j, k) = (1, 2, 3)$. Consider the projection

$$\pi \colon F_4(\mathbb{R}^N) \longrightarrow F_3(\mathbb{R}^N) \tag{1.5}$$

that forgets the fourth point of the configuration. It is a fibration with fiber

$$F = \mathbb{R}^N \setminus \{z_1, z_2, z_3\}.$$

We will obtain β by integration along the fiber of π of some suitable differential form α on $F_4(\mathbb{R}^N)$. To ensure convergence of the integral, we replace the spaces in the fibration (1.5) by their Fulton-MacPherson compactifications C[4] and C[3] so that the fiber becomes diffeomorphic to a closed disk in $\mathbb{R}^N$ with three small open disks removed. We will denote this fiber by $\overline{F}$. Intuitively, each of the three inner boundary spheres of $\overline{F}$ corresponds to points z_4 becoming infinitesimaly close to z_1, z_2, or z_3, (which we denote by $z_4 \simeq z_i$), and the outer boundary sphere of $\overline{F}$ corresponds to the point z_4 going to infinity (which we denote by $z_4 \simeq \infty$)).

Now consider the map

$$\theta := (\theta_{14}, \theta_{24}, \theta_{34}) \colon \mathrm{C}[4] \longrightarrow S^{N-1} \times S^{N-1} \times S^{N-1}. \tag{1.6}$$

The pullback form

$$\theta^*(\mathrm{vol} \times \mathrm{vol} \times \mathrm{vol})$$

is a cocycle in $\Omega_{DR}^{3N-3}(\mathrm{C}[4])$ and is exactly

$$\omega_{14} \wedge \omega_{24} \wedge \omega_{34}.$$

Integration along the fiber of π is a linear map

$$\pi_* = \int_{\overline{F}}: \quad \Omega_{DR}^{3N-3}(\mathrm{C}[4]) \longrightarrow \Omega_{DR}^{2N-3}(\mathrm{C}[3])$$
$$\alpha \longmapsto \int_{\overline{F}} \alpha.$$

The integration takes place along the variable z_4 in the fiber $\overline{F}$ which corresponds to the fourth component of a configuration $z \in \mathrm{C}[4]$. The map π_* satisfies a fiberwise Stokes formula

$$d(\int_{\overline{F}} \alpha) = \int_{\overline{F}} d(\alpha) \pm \int_{\partial\overline{F}} \alpha. \tag{1.7}$$

When $\alpha = \omega_{14} \wedge \omega_{24} \wedge \omega_{34}$, the first term on the right side of (1.7) vanishes because α is a cocyle. We study its second term. One of the boundary components of $\overline{F}$ corresponds to $\{z_4 \simeq z_1\} \subset \partial\overline{F}$, and θ_{14} restricts to a diffeomorphism

$$\theta_{14}\colon \{z_4 \simeq z_1\} \xrightarrow{\cong} S^{N-1}.$$

We then have

$$\int_{\{z_4 \simeq z_1\}} \omega_{14} \wedge \omega_{24} \wedge \omega_{34} = \int_{\{z_4 \simeq z_1\}} \omega_{14} \wedge \omega_{21} \wedge \omega_{31} = \left(\int_{S^{N-1}} \mathrm{vol} \right) \cdot \omega_{21} \wedge \omega_{31} = \omega_{21} \wedge \omega_{31}.$$

Similarly the components corresponding to $z_4 \simeq z_2$ and $z_4 \simeq z_2$ give the two other summands of the 3-term relation (1.3). Another argument shows that the integral along the outer boundary corresponding to $z_4 \simeq \infty$ vanishes. Thus

$$\beta := \int_{\overline{F}} \alpha$$

satisfies Equation (1.4) and is naturally defined.

This algorithm for constructing β can be encoded by a diagram Γ as pictured in Figure 1.2. In this diagram, vertices $1, 2, 3$ (pictured on a line segment) are called *external* and vertex 4 is called *internal*.

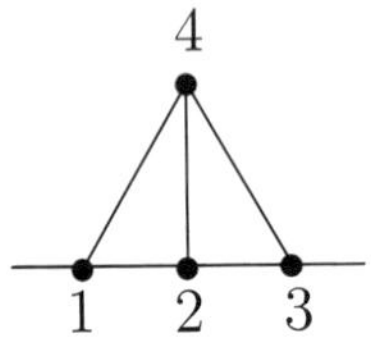

FIGURE 1.2. The diagram Γ that cancels the 3-term relation from Figure 1.1.

The three edges $(1,4)$, $(2,4)$, and $(3,4)$ correspond to the three components of the map θ from (1.6). To such a diagram we have associated the differential form

$$\mathrm{I}(\Gamma) := \int_{\text{fiber}} \theta_{14}^*(\mathrm{vol}) \wedge \theta_{24}^*(\mathrm{vol}) \wedge \theta_{34}^*(\mathrm{vol}) = \pi_*(\theta^*(\underset{3}{\times} \mathrm{vol})) \tag{1.8}$$

where the points of the fiber are those labeled by internal vertices in the diagram Γ (that is, not on the horizontal line, which is z_4 in this case).

We define the coboundary of such a diagram Γ by taking the sum over all possible contractions of an edge with not all endpoints on the line. In particular, for Γ

as in Figure 1.2, its coboundary is exactly the diagrams of Figure 1.1 corresponding to the 3-term relation specialized to $n = 3$ and $(i, j, k) = (1, 2, 3)$. Applying I, defined similarly as in (1.8), to the diagrams of Figure 1.1 gives the right hand side of (1.4), which we have shown to be $d(\mathrm{I}(\Gamma))$. In other words, I commutes with the differential in this example.

The vector space of all such "admissible" diagrams will be denoted by $\mathcal{D}$ and will be endowed with the structure of a cooperad in CDGA. The generalization of Formula (1.8) will define the Kontsevich configuration space integral I from (1.1). An algebraic computation will show that $\mathcal{D}(n)$, where n is the number of external vertices (the ones drawn on the horizontal line segment), is quasi-isomorphic to $\mathrm{H}^*(\mathrm{C}[n])$, from which we will deduce that I in (1.1) is a quasi-isomorphism and hence that $\mathrm{C}[n]$ is formal. Since these quasi-isomorphisms respect the cooperadic structure, this will prove the formality of the operad $\mathrm{C}[\bullet]$ which is equivalent to the little disks operad.

There is one last technical issue. The operad structure on $\mathrm{C}[n]$ corresponds to the inclusions of various faces of the boundary of $\mathrm{C}[n]$. Therefore, in order for I to be a map of cooperads, it is essential that the forms $\mathrm{I}(\Gamma)$ are well-defined on this boundary. However, the projection

$$\pi\colon \mathrm{C}[n+l] \longrightarrow \mathrm{C}[n]$$

is unfortunately not a smooth submersion on the boundary $\partial\,\mathrm{C}[n]$ (see Example 5.9.1), and hence $\mathrm{I}(\Gamma)$ need not be a smooth form on this boundary. To fix this problem we will replace Ω_{DR} by the CDGA Ω_{PA} of PA forms as defined in [**22**, Appendix]. These were studied in great detail in [**18**] and are reviewed in Chapter 4.

1. Plan of the paper

For a faster run through this paper, the reader could, after reading the Introduction, jump directly to the beginning of Chapter 9 to get a better idea of the construction of the quasi-isomorphism of operads

$$\mathrm{I}\colon \mathcal{D}(\bullet) \longrightarrow \Omega_{PA}(\mathrm{C}[\bullet])$$

which is central to our proofs. Along the way, a quick look at Sections 5.1 and 6.1–6.2 will supply a better sense of the Fulton-MacPherson operad $\mathrm{C}[\bullet]$ and the CDGA cooperad of admissible diagrams $\mathcal{D}(\bullet)$, respectively.

The plan of the paper is as follows (see also the Table of Contents at the beginning of the paper).

- In Chapter 2 we fix some notation, and in particular establish some terminology relating to linear orders and weak ordered partitions which will be useful in describing the operad structure maps.
- In Chapter 3 we define in detail what we mean by formality for operads. This is not as straighforward as one might wish because the Sullivan-deRham functor A_{PL} (or its semi-algebraic analog Ω_{PA}) does not turn operads into genuine cooperads of CDGAs. Our definition, however, is practical enough for applications.
- In Chapter 4 we review the functor Ω_{PA} of PA forms. This is the analog for semi-algebraic spaces of the deRham functor Ω_{DR} of differential forms for smooth manifolds. We review the main results we will need from this theory, such as the notion of semi-algebraic chains $\mathrm{C}_*(X)$ on a semi-algebraic set X, which are weakly equivalent to singular chains; the fact that Ω_{PA} encodes (monoidaly) the

real homotopy type of compact semi-algebraic sets; and the important notion of integration along the fiber, or pushforward, of a "minimal" PA form along a semi-algebraic bundle.

- In Chapter 5 we define and study in detail the Fulton-MacPherson operad $\mathrm{C}[\bullet]$ and prove the results about this operad that are necessary for establishing certain properties of the Kontsevich configuration space integral. We also review the fact that the Fulton-MacPherson operad is equivalent to the little balls operad.
- In Chapter 6 we construct the combinatorial CDGA $\mathcal{D}(n)$ of *admissible diagrams* (on n external vertices), built from a larger companion CDGA $\widehat{\mathcal{D}}(n)$ of *diagrams*. The CDGA $\mathcal{D}(n)$ will later be shown to be quasi-isomorphic to both $\Omega_{PA}(\mathrm{C}[n])$ and its cohomology.
- In Chapter 7 we endow first $\widehat{\mathcal{D}}$ and then $\mathcal{D}$ with the structure of a cooperad. The cooperad structure is obtained by considering *condensations*, which will have already appeared in the study of the Fulton-MacPherson operad in Chapter 5.
- In Chapter 8, we prove that the cooperad $\mathcal{D}$ is quasi-isomorphic to the cohomology of the Fulton-MacPherson operad.
- In Chapter 9 we construct the Kontsevich configuration space integrals, which are CDGA maps

$$\widehat{\mathrm{I}}\colon \widehat{\mathcal{D}}(n) \longrightarrow \Omega_{PA}(\mathrm{C}[n]) \quad \text{and} \quad \mathrm{I}\colon \mathcal{D}(n) \longrightarrow \Omega_{PA}(\mathrm{C}[n]).$$

We prove that they are (almost) morphisms of cooperads. The arguments use many properties of the Fulton-MacPherson operad developed in Chapter 5.

- In Chapter 10 we collect the results of the previous two chapters to deduce our main formality results. In particular, we prove that I is a quasi-isomorphism.
- Lastly, for the convenience of the reader we have included an index of notation in the Appendix.

CHAPTER 2

Notation, linear orders, weak partitions, and operads

In this chapter we fix some notation, most of which is standard. We also review the notion of linear orders and introduce the notion of a weak ordered partition which is useful in describing the operad structure maps.

2.1. Notation

$\mathbb{K}$ will be a commutative ring with unit, often $\mathbb{R}$.

An integer $N \geq 1$ (which gives the ambient dimension) will be fixed.

For a set A we denote by $|A|$ its cardinality. We denote by $\mathrm{Perm}(A)$ the group of permutations of A. For a nonnegative integer n, we set $\underline{n} = \{1, \ldots, n\}$. We will sometimes identify n and the set $\underline{n}$. The set of all functions from a set X to a set Y is denoted by Y^X.

When $f\colon X \to Y$ is a map and $A \subset X$, we denote the restriction of f to A by $f|A$.

We denote the one-point space by $*$.

We use the notation $x :=$ def to state that the left hand side is defined by the right hand side.

An extended index of notation is in the Appendix.

2.2. Linear orders

DEFINITION 2.2.1. A *linearly ordered* (or a *totally ordered*) set is a pair $(L, \leq)$ where L is a set and $\leq$ is a reflexive, transitive, and antisymmetric relation on L such for any $x, y \in L$ we have $x \leq y$ or $y \leq x$. We write $x < y$ when $x \leq y$ and $x \neq y$.

Given two disjoint linearly ordered sets $(L_1, \leq_1)$ and $(L_2, \leq_2)$ their *ordered sum* is the linearly ordered set $L_1 \oslash L_2 := (L_1 \cup L_2, \leq)$ such that the restriction of $\leq$ to L_i is the given order $\leq_i$ and such that $x_1 \leq x_2$ when $x_1 \in L_1$ and $x_2 \in L_2$.

More generally if $\{L_p\}_{p \in P}$ is a family of linearly ordered sets indexed by a linearly ordered set P, its *ordered sum*

$$\bigoslash_{p \in P} L_p$$

is the disjoint union $\amalg_{p \in P} L_p$ equipped with a linear order $\leq$ whose restriction to each L_p is the given order on that set and such that $x < y$ when $x \in L_p$ and $y \in L_q$ with $p < q$ in P.

It is clear that the ordered sum $\oslash$ is associative but not commutative.

We define the *position* function on a linearly ordered finite set $(L, \leq)$ as the unique order-preserving isomorphism

$$\mathrm{pos}\colon L \longrightarrow \{1, \ldots, |L|\}.$$

We write $\mathrm{pos}(x : L)$ for $\mathrm{pos}(x)$ when we want to emphasize the underlying ordered set L.

2.3. Weak ordered partitions

The following terminology will be useful in the description of operad structures in the next section.

DEFINITION 2.3.1. A *weak partition* of a finite set A is a map $\nu\colon A \to P$, where P is a finite set. The preimages $\nu^{-1}(p)$, for $p \in P$, are the *elements* of the partition. Since we do not ask ν to be surjective, some of the elements $\nu^{-1}(p)$ can be empty, and hence the adjective *weak*. The weak partition is *degenerate* if ν is not surjective, and *non-degenerate* otherwise. We will simply say *partition* for a non-degenerate weak partition. The (weak) partition ν is *ordered* if its codomain P is equipped with a linear order. The *undiscrete* partition is the partition $\nu\colon A \to \{1\}$ whose only element is A.

2.4. Operads and cooperads

Here we review the definition of operads that we will use. Let $(\mathcal{C}, \otimes, \mathbf{1})$ be a symmetric monoidal category. Let IsoFin be the category whose objects are finite sets (including the empty set) and whose morphisms are bijections between them. This category is equivalent to the category with one object for each integer $n \geq 0$ along with the symmetric group $\Sigma_n = \mathrm{Perm}(\underline{n})$ as its set of automorphisms, and no other morphisms. A *symmetric sequence in* $\mathcal{C}$ is a functor

$$\mathcal{O}\colon \mathrm{IsoFin} \longrightarrow \mathcal{C}.$$

Thus a symmetric sequence in $\mathcal{C}$ is determined by a sequence $(\mathcal{O}(n))_{n\geq 0}$ of objects of $\mathcal{C}$ together with an action of Σ_n on $\mathcal{O}(n)$.

An *operad* $\mathcal{O}$ is a symmetric sequence together with a *unit map*

$$u\colon \mathbf{1} \longrightarrow \mathcal{O}(1)$$

and, for each ordered weak partition $\nu\colon A \to P$, natural *operad structure maps*

$$\Theta_\nu\colon \mathcal{O}(P) \otimes \bigotimes_{p\in P} \mathcal{O}(\nu^{-1}(p)) \longrightarrow \mathcal{O}(A) \tag{2.1}$$

satisfying the usual associativity, unital, and equivariance conditions. Here the monoidal product $\bigotimes_{p\in P}$ is taken of course in the linear order of P.

A *cooperad* is an operad in the opposite category.

Our operads have an object $\mathcal{O}(0) = \mathcal{O}(\emptyset)$ in arity 0. If we were working with operads without a nullary term, then we would only need non-degenerate partitions ν.

When investigating (co)operads, we will often fix the following setting:

SETTING 2.4.1. *Fix an ordered weak partition $\nu\colon A \to P$, with A and P finite, and P linearly ordered. We assume that $0 \notin P$ and set*

$$P^* := \{0\} \oslash P \tag{2.2}$$

where $\oslash$ *is the ordered sum defined in Section* 2.2. *Set* $A_p = \nu^{-1}(p)$ *for* $p \in P$, *and* $A_0 = P$.

Under this setting the structure maps (2.1) become

$$\Theta_\nu : \bigotimes_{p \in P^*} \mathcal{O}(A_p) \longrightarrow \mathcal{O}(A).$$

CHAPTER 3

CDGA models for operads

In this chapter we give precise meaning to the notion of a CDGA model for a topological operad or for a morphism of topological operads. Our definition, although not difficult, is perhaps not so elegant, but it suffices for the applications we have in mind. At the end of the chapter we sketch an alternative, more concise definition.

Recall that Sullivan [**30**] (see [**7**] or [**11**] for a complete development of the theory) constructed a contravariant functor of piecewise polynomial forms over a field $\mathbb{K}$ of characteristic 0,

$$A_{PL}(-;\mathbb{K})\colon \mathrm{Top} \longrightarrow \mathrm{CDGA}$$

which mimics the deRham differential algebra of smooth differential forms on a manifold. Here CDGA is the category of *commutative differential graded* $\mathbb{K}-$*algebras* (or CDGA for short) which are non-negatively graded. Sometimes we will also consider $\mathbb{Z}$*-graded CDGAs* which can be non trivial in negative degree, but those are not the objects of the category CDGA. A CDGA (A,d) is a *CDGA model* (over $\mathbb{K}$) for a space X if the CDGAs (A,d) and $A_{PL}(X;\mathbb{K})$ are weakly equivalent, by which we mean that there exists a chain of quasi-isomorphisms of CDGAs connecting them:

$$(A,d) \xleftarrow{\simeq} \cdots \xrightarrow{\simeq} A_{PL}(X;\mathbb{K}).$$

The main feature of the theory is that when X is a simply-connected topological space with finite Betti numbers and $\mathbb{K}=\mathbb{Q}$, then any CDGA model for X determines the rational homotopy type of X. Moreover, many rational homotopy invariants, like the rational cohomology algebra $\mathrm{H}^*(X;\mathbb{Q})$ or the rational homotopy Lie algebra $\pi_*(\Omega X)\otimes\mathbb{Q}$ can easily be recovered from the model (A,d). For fields $\mathbb{K}$ other than the rationals, we have

$$A_{PL}(-;\mathbb{K}) = A_{PL}(-;\mathbb{Q})\otimes_{\mathbb{Q}}\mathbb{K},$$

and by extension we say that the quasi-isomorphism type of $A_{PL}(X;\mathbb{K})$ determines the $\mathbb{K}$*-homotopy type* of X. We just write $A_{PL}(X)$ when the field $\mathbb{K}$ is understood.

Also, if $f\colon X\to Y$ is a map of spaces, we say that a CDGA morphism

$$\phi\colon (B,d_B)\longrightarrow (A,d_A)$$

is a *CDGA model* for f if there exists a zig-zag of weak equivalences connecting ϕ and $A_{PL}(f;\mathbb{K})$, that is, if there exists a commutative diagram of CDGAs

$$\begin{array}{ccccccccc}
(B,d_B) & \xleftarrow{\simeq} & \bullet & \xrightarrow{\simeq} & \cdots & \xleftarrow{\simeq} & \bullet & \xrightarrow{\simeq} & A_{PL}(Y;\mathbb{K}) \\
\downarrow{\scriptstyle\phi} & & \downarrow & & & & \downarrow & & \downarrow{\scriptstyle A_{PL}(f;\mathbb{K})} \\
(A,d_A) & \xleftarrow{\simeq} & \bullet & \xrightarrow{\simeq} & \cdots & \xleftarrow{\simeq} & \bullet & \xrightarrow{\simeq} & A_{PL}(X;\mathbb{K})
\end{array}$$

in which the horizontal arrows are quasi-isomorphisms.

We would like to define a similar notion of a CDGA model for a topological operad $\mathcal{O}$. A naive definition would be that such a model is a cooperad $\mathcal{A}$ of CDGAs that is connected by weak equivalences of CDGA cooperads to $A_{PL}(\mathcal{O})$. However, there is a problem with this definition because the contravariant functor A_{PL} is not comonoidal as there is no suitable natural map

$$A_{PL}(X \times Y) \longrightarrow A_{PL}(X) \otimes A_{PL}(Y). \tag{3.1}$$

Therefore it seems that there is no cooperad structure on $A_{PL}(\mathcal{O})$ naturally induced from the operad structure on $\mathcal{O}$. On the other hand, A_{PL} is monoidal through the Kunneth quasi-isomorphism

$$\kappa\colon A_{PL}(X) \otimes A_{PL}(Y) \xrightarrow{\simeq} A_{PL}(X \times Y). \tag{3.2}$$

This morphism becomes an isomorphism in the homotopy category, and its inverse should correspond to the homotopy class of the missing map (3.1). We would thus like to say that $A_{PL}(\mathcal{O})$ is a cooperad "up to homotopy". However, this sort of "up to homotopy" structure needs to be handled with more care than is necessary for our purpose, and so we will not pursue this in detail here and will just give an indication of such a notion at the end of the chapter. Instead we will propose in Definition 3.1 an ad hoc definition of a CDGA model for an operad.

There is a second difficulty which we will have do deal with and which comes from the proof of the formality itself. Namely, in Kontsevich's proof of the weak equivalence between the (up to homotopy) cooperad $A_{PL}(\mathcal{B})$ and its cohomology, a functor Ω_{PA} (to be reviewed in Chapter 4) is used. This functor is weakly equivalent to $A_{PL}(-;\mathbb{R})$ but is defined only after restriction to a subcategory of Top, namely the category of compact semi-algebraic sets. This is analogous to the fact that the deRham CDGA of smooth differential forms Ω_{DR} is weakly equivalent to $A_{PL}(-;\mathbb{R})$ after restriction to the subcategory of smooth manifolds. Consequently, our modeling functors will sometimes be defined on some subcategory $u\colon \mathcal{T} \hookrightarrow \mathrm{Top}$.

To finally define our notion of a CDGA model for an operad, we will need a few definitions.

Two cooperads of CDGAs, $\mathcal{A}$ and $\mathcal{A}'$, are *weakly equivalent* if they are connected by a chain of quasi-isomorphism of CDGA cooperads,

$$\mathcal{A} \xleftarrow{\simeq} \cdots \xrightarrow{\simeq} \mathcal{A}'.$$

Let $(\mathcal{T}, \times, \mathbf{1})$ be a symmetric monoidal category and let

$$u\colon \mathcal{T} \longrightarrow \mathrm{Top}$$

be a symmetric *strongly* monoidal covariant functor, where by strongly we mean that the natural map

$$u(X) \times u(Y) \xrightarrow{\cong} u(X \times Y) \tag{3.3}$$

is an isomorphism and $u(\mathbf{1}) = *$ is the one-point space.

For us, a contravariant functor

$$\Omega\colon \mathcal{T} \longrightarrow \mathrm{CDGA}$$

is *symmetric monoidal* if it is equipped with a natural map

$$\kappa\colon \Omega(X) \otimes \Omega(Y) \longrightarrow \Omega(X \times Y) \tag{3.4}$$

satisfying the usual axioms and such that $\Omega(\mathbf{1}) = \mathbb{K}$. In particular $A_{PL} \circ u$ is symmetric monoidal.

A *natural monoidal quasi-isomorphism* between two such contravariant symmetric monoidal functors Ω and Ω' is a natural transformation

$$\theta \colon \Omega \longrightarrow \Omega'$$

that induces an isomorphism in homology and that commutes with the monoidal structure maps. Two symmetric monoidal contravariant functors are *weakly equivalent* if they are connected by a chain of natural monoidal quasi-isomorphisms. If Ω is weakly equivalent to $A_{PL} \circ u$ then the morphism κ of (3.4) is a quasi-isomorphism because the corresponding one for A_{PL} in (3.2) is as well and because of the isomorphism (3.3).

Our definition of CDGA models for cooperads is then

DEFINITION 3.1. A CDGA cooperad $\mathcal{A}$ is a *CDGA model* for a topological operad $\mathcal{O}$ if there exist

- a CDGA cooperad $\mathcal{A}'$ weakly equivalent to $\mathcal{A}$;
- a symmetric monoidal category $(\mathcal{T}, \times, \mathbf{1})$;
- a symmetric strongly monoidal covariant functor $u \colon \mathcal{T} \to \mathrm{Top}$;
- an operad $\mathcal{O}'$ in $\mathcal{T}$ such that $u(\mathcal{O}')$ is weakly equivalent to $\mathcal{O}$;
- a symmetric monoidal contravariant functor Ω weakly equivalent to $A_{PL} \circ u$;
- for each $n \geq 0$ a Σ_n-equivariant quasi-isomorphism

  $$J_n \colon \mathcal{A}'(n) \xrightarrow{\simeq} \Omega(\mathcal{O}'(n))$$

  such that, for each $k \geq 0$ and $n_1, \ldots, n_k \geq 0$ with $n = n_1 + \cdots + n_k$, the following diagram commutes:

  $$\begin{array}{ccc}
  \mathcal{A}'(n) & \xrightarrow[\simeq]{J_n} & \Omega(\mathcal{O}'(n)) \\
  & & \downarrow{\scriptstyle \Omega(\Phi)} \\
  \downarrow{\scriptstyle \Psi} & & \Omega(\mathcal{O}'(k) \times \mathcal{O}'(n_1) \times \cdots \times \mathcal{O}'(n_k)) \\
  & & \uparrow{\scriptstyle \kappa}\ \simeq \\
  \mathcal{A}'(k) \otimes \mathcal{A}'(n_1) \otimes \ldots \mathcal{A}'(n_k) & \xrightarrow[J_k \otimes J_{n_1} \otimes \cdots \otimes J_{n_k}]{\simeq} & \Omega(\mathcal{O}'(k)) \otimes \Omega(\mathcal{O}'(n_1)) \otimes \cdots \otimes \Omega(\mathcal{O}'(n_k)).
  \end{array}$$

  Here Ψ and Φ are the (co)operad structure maps on $\mathcal{A}'$ and $\mathcal{O}'$ respectively, and the composition

  $$\mathcal{A}'(1) \xrightarrow{J_1} \Omega(\mathcal{O}'(1)) \xrightarrow{\Omega(\eta)} \Omega(\mathbf{1}) \cong \mathbb{K}$$

  is required to be the counit of $\mathcal{A}'$, where η is the unit of $\mathcal{O}'$.

If κ was an isomorphism, then $\kappa^{-1} \circ \Omega(\Phi)$ would define a cooperad structure on $\Omega(\mathcal{O}')$ and the above diagram would simply mean that the cooperads $\mathcal{A}'$ and $\Omega(\mathcal{O}')$ are weakly equivalent.

The main examples of the above that we will consider are:

- the category $\mathcal{T} = \mathrm{CompactSemiAlg}$ of compact semi-algebraic sets (Chapter 4);

- the forgetful functor $u\colon \text{CompactSemiAlg} \to \text{Top}$;
- the functor $\Omega = \Omega_{PA}$ of semi-algebraic forms (Chapter 4);
- the topological operad of little balls $\mathcal{O} = \mathcal{B}_N$;
- the Fulton-MacPherson semi-algebraic operad $\mathcal{O}' = \mathrm{C}[\bullet]$ (Chapter 5), and
- its cohomology $\mathcal{A} = \mathrm{H}^*(\mathrm{C}[\bullet])$;
- the cooperad of admissible diagrams $\mathcal{A}' = \mathcal{D}$ (Chapters 6-7); and
- the Kontsevich configuration space integral $J_n = \mathrm{I}\colon \mathcal{D}(n) \to \Omega_{PA}(\mathrm{C}[n])$ (Chapter 9).

We will let the reader generalize Definition 3.1 in an obvious way to say when a morphism of CDGA cooperads

$$\phi\colon \mathcal{A}_1 \longrightarrow \mathcal{A}_2$$

is a *CDGA model for a morphism of topological operads*

$$f\colon \mathcal{O}_2 \longrightarrow \mathcal{O}_1.$$

DEFINITION 3.2. A topological operad is *formal* over $\mathbb{K}$ if the induced cohomology algebra cooperad is a CDGA model for this operad over $\mathbb{K}$.

A morphism of topological operads is *formal* if the induced morphism in cohomology is a CDGA model for this operad morphism.

This definition, albeit perhaps a bit ad hoc, is good enough for the applications we have in mind. A more elegant definition would have to use a precise notion of a (co)operad up to homotopy as follows.

Recall, for example from [**15**, §1.2], that an operad can be seen as a functor on the category of trees. More precisely let Tree be the category whose objects are rooted planar trees and morphisms compositions of contractions of non-terminal edges. Given trees $S, T_1, \ldots, T_k$ where S has k leaves and each T_i has n_i leaves, one can build a new tree $S(T_1, \ldots, T_k)$ with $n_1 + \cdots + n_k$ leaves by grafting the root of each tree T_i to the corresponding leaf of S. For $n \geq 0$ we denote by $\langle n \rangle$ the tree with n leaves and no internal vertex, that is a tree which is indecomposable with respect to the grafting operation. Then an operad $\mathcal{O}$ in a symmetric monoidal category $\mathcal{C}$ can be seen as a functor

$$O\colon \text{Tree} \longrightarrow \mathcal{C}$$

where $O(\langle n \rangle) = \mathcal{O}(n)$, for $n \geq 0$. In order for a functor O to define an operad one asks for isomorphisms

$$\alpha_{(S,T_1,\ldots,T_k)}\colon O(S(T_1, \ldots, T_k)) \xrightarrow{\cong} O(S) \otimes \otimes_{i=1}^{k} O(T_i)$$

satisfying obvious associativity, unital, and equivariance relations.

There is a morphism in Tree given by

$$\langle k \rangle(\langle n_1 \rangle, \ldots, \langle n_k \rangle) \longrightarrow \langle n_1 + \cdots + n_k \rangle$$

and its image under the functor O composed with the inverse of the isomorphism α gives the structure maps of the operad.

An *operad up to homotopy* is an analogous functor O except that one only asks $\alpha_{(S,T_1,\ldots,T_k)}$ to be a weak equivalence instead of an isomorphism. Similarly we can define cooperads up to homotopy.

If $\mathcal{O}$ is a topological operad, then $A_{PL}(\mathcal{O})$ naturally becomes a cooperad up to homotopy in this sense, with the weak equivalences α constructed from the Kunneth quasi-isomorphism (3.2). There is also an obvious notion of morphisms of (co)operads up to homotopy and of weak equivalences. One could check that if a CDGA cooperad $\mathcal{A}$ is a CDGA model for a topological operad $\mathcal{O}$ in the sense of Definition 3.1, then $\mathcal{A}$ and $A_{PL}(\mathcal{O})$ are also weakly equivalent as cooperads up to homotopy. This therefore might give a better definition of an operad model, but it is possible that some further "∞-version" would be necessary for obtaining something useful.

CHAPTER 4

Real homotopy theory of semi-algebraic sets

In this chapter we give a brief review of Kontsevich and Soibelman's theory of semi-algebraic differential forms which is outlined in [**22**, §8]. In particular we discuss the functor Ω_{PA} which is analogous to the deRham functor Ω_{DR} for smooth manifolds. That functor and the way it encodes real homotopy theory of semi-algebraic sets was developed in full detail by the authors jointly with Robert Hardt and Victor Turchin in [**18**].

DEFINITION 4.1 ([**5**]). A *semi-algebraic set* is a subset of $\mathbb{R}^p$ that is obtained by finite unions, finite intersections, and complements of subsets defined by polynomial equations and inequalities. A *semi-algebraic map* is a continuous map between semi-algebraic sets whose graph is a semi-algebraic set.

We will consider the categories SemiAlg (and CompactSemiAlg) of (compact) semi-algebraic sets. Endowed with the cartesian product, this category becomes symmetric monoidal and the obvious forgetful functor

$$u\colon \mathrm{SemiAlg} \longrightarrow \mathrm{Top}$$

is strongly symmetric monoidal because of the natural homeomorphism

$$u(X) \times u(Y) \xrightarrow{\cong} u(X \times Y).$$

We have for a semi-algebraic set X a functorial chain complex of *semi-algebraic chains* $\mathrm{C}_*(X)$ [**18**, Definition 3.1], which is weakly equivalent to singular chains. A typical element of $\mathrm{C}_k(X)$ is represented by a semi-algebraic map $g\colon M \to X$ from a semi-algebraic compact oriented manifold M of dimension k. This element is denoted by $g_*([\![M]\!]) \in \mathrm{C}_k(X)$. In particular, taking $g = \mathrm{id}_M$,

$$[\![M]\!] \in \mathrm{C}_k(M) \tag{4.1}$$

represents a canonically defined fundamental class of the manifold M at the level of semi-algebraic chains. Also, a semi-algebraic map $f\colon X \to Y$ induces a chain map

$$f_*\colon \mathrm{C}_*(X) \longrightarrow \mathrm{C}_*(Y). \tag{4.2}$$

We in addition have a contravariant functor of *minimal forms* [**18**, Section 5.2]

$$\Omega_{\min}\colon \mathrm{SemiAlg} \longrightarrow \mathrm{CDGA}\,. \tag{4.3}$$

By definition, a minimal form of degree k on X is represented by a linear combination of

$$\mu = f_0 \cdot df_1 \wedge \cdots \wedge df_k$$

where

$$f_0, f_1, \ldots, f_k\colon X \longrightarrow \mathbb{R}$$

are semi-algebraic maps. Even though the f_i's may not be everywhere smooth, we can define a differential $d\mu$ which is again a minimal form. Also for a compact semi-algebraic oriented manifold M of dimension k and a semi-algebraic map $g\colon M \to X$, we can evaluate the form μ on $g_*([\![M]\!]) \in \mathrm{C}_k(X)$ by the formula

$$\langle \mu\,,\, g_*[\![M]\!]\rangle := \int_M g^*(f_0 \cdot df_1 \wedge \cdots \wedge df_k). \tag{4.4}$$

The convergence of the integral on the right is a consequence of the semi-algebraicity of M. Indeed that integral is the same as

$$\int_{f_*(g_*(M))} x_0 \cdot dx_1 \wedge \cdots \wedge dx_k \tag{4.5}$$

where $f_*(g_*(M))$ is the image of M in $\mathbb{R}^{k+1}$ (counted with multiplities) under the composition of g and $f := (f_0, f_1, \ldots, f_k)$. Thus $f_*(g_*(M))$ is a compact semi algebraic-set of dimension $\leq k$, which implies that its k-volume is finite (this would not be true for non semi-algebraic compact sets.) Hence the integral in Equation (4.5) converges. See [**18**, Theorem 2.4 and beginning of Section 3] for more details.

In this paper, the only minimal forms that we will use are the standard volume form on the sphere and its pullbacks along semi-algebraic maps.

The CDGA of minimal forms embeds in that of *PA forms* [**18**, Section 5.4] ("PA" stands for "piecewise algebraic")

$$\Omega_{PA}\colon \mathrm{SemiAlg} \longrightarrow \mathrm{CDGA}\,. \tag{4.6}$$

Roughly speaking, PA forms are obtained by integration along the fiber of minimal forms along oriented semi-algebraic bundles, which are recalled below. The important feature is the following

THEOREM 4.2 ([**18**, Theorem 7.1]). *When restricted to the category of compact semi-algebraic sets, the contravariant symmetric monoidal functors Ω_{PA} and $A_{PL}(u(-);\mathbb{R})$ are weakly equivalent.*

Another important feature of minimal and PA forms is that classical integration along the fiber for smooth forms can be extended to the semi-algebraic framework. To explain, we have from [**18**, Section 8] the notion of a *semi-algebraic bundle*, or *SA bundle* for short, which is the obvious generalization of the usual definition of a locally trivial bundle.

An SA bundle

$$\pi\colon E \longrightarrow B$$

is *oriented* if its fibers are compact oriented semi-algebraic manifolds, with orientation which is locally constant in an obvious sense. For each $b \in B$ we then have the fundamental class of the fiber over b,

$$[\![\pi^{-1}(b)]\!] \in \mathrm{C}_k(\pi^{-1}(b)), \tag{4.7}$$

where k is the dimension of the fiber.

Given an oriented SA bundle $\pi\colon E \to B$ whose fibers are compact SA manifolds, there exists a subbundle

$$\pi^\partial\colon E^\partial \to B \tag{4.8}$$

whose fibers are the boundaries of the fibers of π. This subbundle is called the *fiberwise boundary* of π (see [**18**, Definition 8.1]). An example is the map

$$\mathrm{proj}_1\colon E := [0,1] \times [0,1] \to [0,1]$$

which projects onto the first factor. In this case the fiberwise boundary is $E^\partial = [0,1] \times \{0,1\}$, but this is not the boundary of E.

For an oriented SA bundle with k-dimensional fiber, there is a linear map of degree $-k$ [**18**, Definition 8.3],

$$\pi_*\colon \Omega^{*+k}_{\mathrm{min}}(E) \longrightarrow \Omega^*_{PA}(B), \tag{4.9}$$

which correponds to integration along the fiber, also called *pushforward.* In some sense PA forms are obtained as (generalized) pushforwards of minimal forms [**18**, Definition 5.20]. Properties of the pushforward that we will need here are collected in [**18**, Section 8.2]. They are analogous to the standard properties of integration of smooth differential forms along the compact fiber of a smooth bundle. In particular one has a fiberwise Stokes formula which we will need later.

CHAPTER 5

The Fulton-MacPherson operad

Fix $N \geq 1$. In this chapter we review the Fulton-MacPherson operad

$$\mathrm{C}[\bullet] = \{\mathrm{C}[n]\}_{n\geq 0}$$

which is weakly equivalent to the little N-disks operad. As a space, each $\mathrm{C}[n]$ is a compactification of the space $\mathrm{C}(n)$ of normalized ordered configurations of n points in $\mathbb{R}^N$. It is a compact semi-algebraic manifold with boundary, and so its real homotopy type is encoded by the semi-algebraic analog of deRham theory, $\Omega_{PA}(\mathrm{C}[n])$. The operad structure maps correspond essentially to inclusions of various faces of the boundary $\partial\,\mathrm{C}[n]$.

We will also in this chapter study *canonical projections*

$$\pi\colon\ \mathrm{C}[n+l] \longrightarrow \mathrm{C}[n]$$

given by forgetting some points of the configuration and will prove that they are SA bundles with compact manifolds as fibers. This fact will be used in Section 9.1 to construct the Kontsevich configuration space integral $\mathrm{I}\colon \mathcal{D}(n) \to \Omega_{PA}(\mathrm{C}[n])$ of (1.1) via integration along the fiber of π. We will also study the interaction of these canonical projections with the operad structure in order to later prove that I is a map of cooperads.

The plan of this chapter is the following:

- 5.1: We define the compactification $\mathrm{C}[n]$, compute its dimension, and characterize its boundary.
- 5.2: We describe the operad structure on $\{\mathrm{C}[n]\}_{n\geq 0}$ and recall that this operad is equivalent to the operad of little balls.
- 5.3: We study the canonical projection $\pi\colon\ \mathrm{C}[n+l] \to \mathrm{C}[n]$ and state that it defines a bundle whose fibers are oriented compact manifolds.
- 5.4: We decompose the boundary $\partial\,\mathrm{C}[n]$ into faces which are images of the $\circ_i$ ("circle-i") operad maps.
- 5.5: We construct *singular* configuration spaces which are variations of spaces $\mathrm{C}[n]$ and will be needed for some technical points.
- 5.6: In this (long) section, we investigate the pullback of a canonical projection along an operad structure map. This will be needed for proving that the Kontsevich configuration space integral respects the (co)operadic structures. We introduce at the beginning of this section the notion of a *condensation* which will also be needed for the definition of the cooperad structure on the space of diagrams in Chapter 7.
- 5.7: We describe a decomposition of the fiberwise boundary of the total space $\mathrm{C}[n+l]$ of a canonical projection. This will be used in proving that Kontsevich's configuration space integral is a chain map.

5.8: We fix an orientation of C$[n]$; this will be important when we integrate forms over this manifold.

5.9: We prove Theorem 5.3.2, stated in Section 5.3, which asserts that the canonical projections are oriented SA bundles. This section also contains an example showing that the canonical projections are not smooth bundles.

On a first pass of this chapter, the reader may just concentrate on Sections 5.1-5.4 to acquire a good sense of the Fulton-McPherson operad. The last five sections are more technical and are needed only for the details of the proof of certain properties of the Kontsevich configuration space integral in Chapter 9. However, the reader should still look at Definition 5.6.1 of a condensation in Section 5.6, as this will be needed in Chapter 7 to define the cooperadic structure on the spaces of diagrams $\widehat{\mathcal{D}}(n)$.

5.1. Compactification of configuration spaces in $\mathbb{R}^N$

We first recall the Fulton-MacPherson compactification C$[n]$ of the configuration space C(n) of n points in $\mathbb{R}^N$. This compactification (or at least some variation of it) was defined in [**12**], with the operad structure given in [**14**], and alternatively by Kontsevich in [**20**, Definition 12] and [**21**, Section 5.1]. We follow Kontsevich's approach, which was corrected by Gaiffi in [**13**, Section 6.2] and developed in detail by Sinha in [**27**] (the equivalence of the Kontsevich and the Fulton-MacPherson definitions follows from Sinha's work as well).

Let A be a finite set of cardinality n which will serve as a set of labels for the points of the configurations. Consider the space

$$\mathrm{Inj}(A,\mathbb{R}^N) := \{x\colon A \hookrightarrow \mathbb{R}^N\} \tag{5.1}$$

of all injective maps from A to $\mathbb{R}^N$. An element $x \in \mathrm{Inj}(A,\mathbb{R}^N)$ is an (ordered) configuration $(x(a))_{a\in A}$ of n distinct points in $\mathbb{R}^N$. This space is topologized as a subspace of the product $(\mathbb{R}^N)^A = \prod_{a\in A}\mathbb{R}^N$.

The space $\mathrm{Inj}(A,\mathbb{R}^N)$ is a smooth open manifold of dimension $N\cdot|A|$. The group of orientation-preserving similarities $\mathbb{R}^N \rtimes \mathbb{R}_0^+$ acts by translation and positive dilation on $\mathbb{R}^N$, and hence diagonally on $\mathrm{Inj}(A,\mathbb{R}^N)$. We denote its orbit space by

$$\mathrm{C}(A) := \mathrm{Inj}(A,\mathbb{R}^N)/(\mathbb{R}^N \rtimes \mathbb{R}_0^+). \tag{5.2}$$

(This space is denoted by $\widetilde{\mathrm{C}_n}(\mathbb{R}^N)$ in [**27**, Definition 3.9].)

When $|A|\geq 2$ the action is free and smooth and hence C(A) is a manifold of dimension

$$\dim \mathrm{C}(A) = N\cdot|A| - N - 1,$$

and when $|A|\leq 1$ then $C(A)$ is a one-point space because the action is transitive.

Define the *barycenter* of a map $x\colon A\to\mathbb{R}^N$ as the point

$$\mathrm{barycenter}(x) = \mathrm{barycenter}(x(a) : a\in A) := \frac{1}{|A|}\sum_{a\in A} x(a) \tag{5.3}$$

and its *radius* as the real number

$$\mathrm{radius}(x) = \mathrm{radius}(x(a) : a\in A) := \max(\|x(a) - \mathrm{barycenter}(x)\| : a\in A). \tag{5.4}$$

When $|A|\geq 2$, C(A) is homeomorphic to the space of *normalized configurations*

$$\mathrm{Inj}_0^1(A,\mathbb{R}^N) := \left\{x\in \mathrm{Inj}(A,\mathbb{R}^N) : \mathrm{barycenter}(x) = 0 \text{ and } \mathrm{radius}(x) = 1\right\}. \tag{5.5}$$

We will use $\mathrm{C}(A)$ and $\mathrm{Inj}_0^1(A, \mathbb{R}^N)$ interchangeably. Most of the time in this paper, a configuration will be denoted by x or y (maybe with some decoration) and, when seen as an element of $\mathrm{Inj}_0^1(A, \mathbb{R}^N)$, its components will be points $x(a)$ for a an element of the set of labels of the components, A.

Denote by S^{N-1} the unit sphere in $\mathbb{R}^N$. Given two distinct elements $a, b \in A$, consider the map

$$\begin{aligned} \theta_{a,b} \colon \mathrm{C}(A) &\longrightarrow S^{N-1} \\ x &\longmapsto \frac{x(b)-x(a)}{\|x(b)-x(a)\|} \end{aligned} \tag{5.6}$$

which gives the direction between two points of the configuration.

For three distinct elements $a, b, c \in A$, also define

$$\begin{aligned} \delta_{a,b,c} \colon \mathrm{C}(A) &\longrightarrow [0, +\infty] \\ x &\longmapsto \frac{\|x(a)-x(b)\|}{\|x(a)-x(c)\|} \end{aligned} \tag{5.7}$$

which gives the relative distance of 3 points of a configuration.

Set

$$\begin{aligned} A^{\{2\}} &= \{(a,b) \in A \times A : a \neq b\} \\ A^{\{3\}} &= \{(a,b,c) \in A \times A \times A : a \neq b \neq c \neq a\} \end{aligned}$$

and consider the map

$$\begin{aligned} \iota \colon \mathrm{C}(A) &\longrightarrow (S^{N-1})^{A^{\{2\}}} \times [0, +\infty]^{A^{\{3\}}} \\ x &\longmapsto \Big((\theta_{a,b}(x))_{(a,b)\in A^{\{2\}}} \,,\, (\delta_{a,b,c}(x))_{(a,b,c)\in A^{\{3\}}} \Big). \end{aligned}$$

Up to translation and dilation, any configuration $x \colon A \hookrightarrow \mathbb{R}^N$ can be recovered from the directions $\theta_{a,b}(x)$ and relative distances $\delta_{a,b,c}(x)$. Hence ι is a homeomorphism onto its image [**27**, Lemma 3.18] and we will identify $\mathrm{C}(A)$ with $\iota(\mathrm{C}(A))$.

DEFINITION 5.1.1. The *Fulton-MacPherson compactification* $\mathrm{C}[A]$ *of* $\mathrm{C}(A)$ is the topological closure of the image of ι, that is,

$$\mathrm{C}[A] := \overline{\iota(\mathrm{C}(A))}.$$

Intuitively, one should think of $x \in \mathrm{C}[A]$ as a "virtual" configuration in which some points are possibly infinitesimally close to each other in such a way that the direction between any two points and the relative distance between three points is always well-defined. These directions and relative distances are given by the maps $\theta_{a,b}$ and $\delta_{a,b,c}$, which obviously extend to $\mathrm{C}[A]$. Moreover an element $x \in \mathrm{C}[A]$ is completely characterized by the values $\theta_{a,b}(x) \in S^{N-1}$ and $\delta_{a,b,c}(x) \in [0, +\infty]$, for distinct $a, b, c \in A$. By abuse of terminology an element $x \in \mathrm{C}[A]$ will be called a configuration and we will talk informally of its components $x(a) \in \mathbb{R}^N$, for $a \in A$.

The following notation will be useful: For a, b, c distinct in A and $x \in \mathrm{C}[A]$, when $\delta_{a,b,c}(x) = 0$ we write

$$x(a) \simeq x(b) \operatorname{rel} x(c). \tag{5.8}$$

This happens exactly when the points $x(a)$ and $x(b)$ are infinitesimaly close to each other in comparison to their distance to $x(c)$. Pictorial interpretations of this situation are given below in Example 5.2.1. In particular Figure 5.2 represents a configuration $x \in \mathrm{C}[6]$ with $N = 2$.

The space $C(A) \subset (\mathbb{R}^N)^A$ and the map ι are clearly semi-algebraic, therefore so is the closure $C[A]$. Moreover, by [**6**] or [**27**], $\mathrm{C}[A]$ is a compact manifold with corners. It is easy to see that the atlases given in those papers are semi-algebraic, and hence $\mathrm{C}[A]$ is a compact semi-algebraic manifold with boundary (charts are given in Lemma 5.9.3).

In conclusion, we have

PROPOSITION 5.1.2. *For a finite set A, $\mathrm{C}[A]$ is a compact semi-algebraic manifold with interior $\mathrm{C}(A)$ and its dimension is given by*

$$\dim(\mathrm{C}[A]) = \begin{cases} 0 & \text{if } |A| \leq 1; \\ N \cdot |A| - N - 1 & \text{if } |A| \geq 2. \end{cases}$$

We also have the following important characterization of the boundary

PROPOSITION 5.1.3. *For $x \in \mathrm{C}[A]$, the following are equivalent conditions:*

$$x \in \partial\, \mathrm{C}[A] \quad \Longleftrightarrow \quad (\exists\, a, b, c \in A \text{ distinct } : x(a) \simeq x(b) \operatorname{rel} x(c)).$$

For $|A| \leq 1$, $\mathrm{C}[A]$ is a one-point space; for $|A| = 2$, it is homeomorphic to the sphere S^{N-1}. For $n \geq 0$ we set $\mathrm{C}[n] := \mathrm{C}[\{1, \ldots, n\}]$.

5.2. The operad structure

We will now define the structure of an operad on

$$\mathrm{C}[\bullet] = \{\mathrm{C}[n]\}_{n \geq 0}.$$

Recall from Chapter 2 the notion of weak ordered partitions and how operad structure maps are associated to them.

Fix a finite set A, a linearly ordered finite set P, and a weak ordered partition $\nu \colon A \to P$. Set

$$P^* = \{0\} \oslash P, \quad A_p = \nu^{-1}(p), \quad \text{and } A_0 = P$$

as in the setting 2.4.1 from Chapter 2. Hence

$$\prod_{p \in P^*} \mathrm{C}[A_p] = \mathrm{C}[P] \times \prod_{p \in P} \mathrm{C}[\nu^{-1}(p)].$$

We now construct an operad structure map

$$\Phi_\nu \colon \prod_{p \in P^*} \mathrm{C}[A_p] \longrightarrow \mathrm{C}[A] \tag{5.9}$$

as follows. Intuitively the configuration $x = \Phi_\nu((x_p)_{p \in P^*})$ is obtained by replacing, for each $p \in P$, the p-th component $x_0(p)$ of the configuration $x_0 \in \mathrm{C}[P]$ by the configuration $x_p \in \mathrm{C}[A_p]$ made infinitesimal. To illustrate, we first give an example.

EXAMPLE 5.2.1. Consider $P = \{\alpha, \beta, \gamma, \delta\}$ (with the linear order $\alpha < \beta < \gamma < \delta$), $A = \{1, 2, 3, 4, 5, 6\}$, and let $\nu \colon A \to P$ be given by

$$\nu(a) = \begin{cases} \alpha, & \text{for } a = 1, 2; \\ \beta, & \text{for } a = 3, 4, 5; \\ \delta, & \text{for } a = 6. \end{cases}$$

Consider

$$x_0 \in C[P] \cong C[4];$$
$$x_\alpha \in C[\{1,2\}] \cong C[2];$$
$$x_\beta \in C[\{3,4,5\}] \cong C[3];$$
$$x_\gamma \in C[\emptyset] \cong C[0] = *;$$
$$x_\delta \in C[\{6\}] \cong C[1] = *$$

and suppose that these configurations are for example as in Figure 5.1 (with $N = 2$).

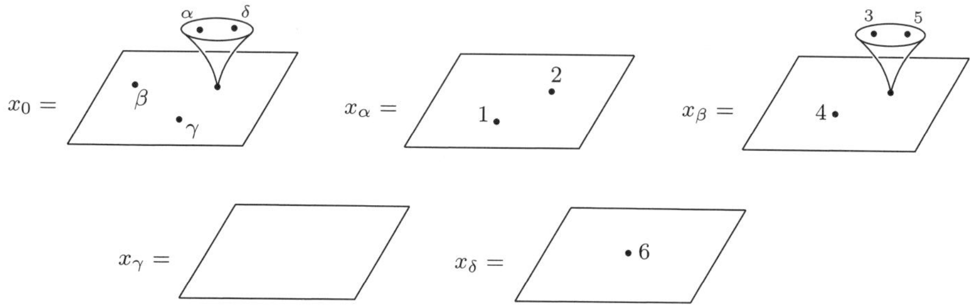

FIGURE 5.1

This kind of pictorial representation of compactified configuration spaces first appeared in [**27**]. The plane represents $\mathbb{R}^N$ and the "funnels" represent infinitesimal configurations. Thus for example, in the picture of x_0, points labeled by α and δ are infinitesimally close to each other from the point of view of β and γ. In notation of relation (5.8), $x_0(\alpha) \simeq x_0(\delta) \operatorname{rel} x_0(\beta)$ and $x_0(\alpha) \simeq x_0(\delta) \operatorname{rel} x_0(\gamma)$. Similarly in the picture of x in Figure 5.2 below, points (labeled by) 4, 3, and 5 are infinitesimally close to each other from the point of view of 6, 1, and 2, but 3 and 5 are infinitesimally close to each other from the point of view of 4, as are 1 and 2 from the point of view of 6.

Then the configuration $x = \Psi_\nu(x_0, x_\alpha, x_\beta, x_\gamma, x_\delta)$ can be represented as in Figure 5.2.

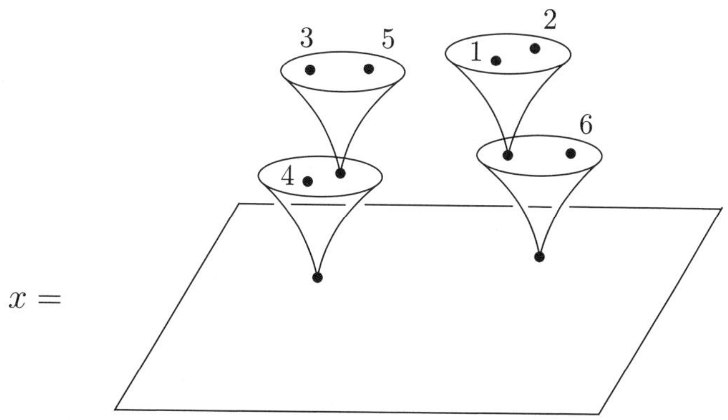

FIGURE 5.2

More precisely, $x = \Phi_\nu((x_p)_{p\in P^*}) \in \mathrm{C}[A]$ is characterized by, for distinct $a, b, c \in A$,

$$\theta_{a,b}(x) = \begin{cases} \theta_{a,b}(x_p), & \text{if } a, b \in A_p \text{ for some } p \in P, \text{ that is } \nu(a) = \nu(b) = p; \\ \theta_{\nu(a),\nu(b)}(x_0), & \text{if } \nu(a) \neq \nu(b), \end{cases}$$

and

$$\delta_{a,b,c}(x) = \begin{cases} \delta_{a,b,c}(x_p), & \text{if } a, b, c \in A_p \text{ for some } p \in P; \\ \delta_{\nu(a),\nu(b),\nu(c)}(x_0), & \text{if } \nu(a),\ \nu(b), \text{ and } \nu(c) \text{ are all distinct}; \\ 0, & \text{if } \nu(a) = \nu(b) \neq \nu(c); \\ 1, & \text{if } \nu(a) \neq \nu(b) = \nu(c); \\ +\infty, & \text{if } \nu(a) = \nu(c) \neq \nu(b). \end{cases}$$

There is an obvious action of the group $\mathrm{Perm}(A)$ of permutations of the set A on $\mathrm{C}[A]$, and in particular of the symmetric group Σ_n on $\mathrm{C}[n]$. We define the unit in $\mathrm{C}[1]$ as its unique point (or more precisely the unique map $u\colon * \to \mathrm{C}[1]$).

The following is straightforward to check (see for example [**28**, Section 4]).

PROPOSITION 5.2.2. *The above data endows* $\mathrm{C}[\bullet] = \{\mathrm{C}[n]\}_{n\geq 0}$ *with the structure of an operad of compact semi-algebraic sets.*

The relevance of the Fulton-MacPherson operad for us is that it is weakly equivalent to the little balls operad, as proved by P. Salvatore:

PROPOSITION 5.2.3. [**25**, Proposition 4.9] *The Fulton-MacPherson operad* $\mathrm{C}[\bullet]$ *of configurations in* $\mathbb{R}^N$ *and the little N-disks operad* $\mathcal{B}$ *are weakly equivalent as topological operads.*

For the sake of keeping this paper as self-contained as possible, we summarize Salvatore's proof here.

SUMMARY OF PROOF OF PROPOSITION 5.2.3. Recall the W construction of Boardman-Vogt [**4**] which associates to a topological operad $\mathcal{O}(\bullet)$ another operad $W\mathcal{O}$ consisting of planar rooted trees τ whose internal edges have length between 0 and 1 and whose internal vertices of valence $i+1$ are decorated by an element of $\mathcal{O}(i)$. The operad $W\mathcal{O}$ is a cofibrant replacement of $\mathcal{O}$. The main idea of the proof is then to construct a map $R\colon W\mathcal{B} \to \mathrm{C}[\bullet]$ that sends a decorated tree τ to the configuration of the centers of the configuration of balls obtained by multicomposition of all the configurations of balls associated to the vertices of τ (after rescaling the configuration of balls at each internal vertex in a way that depends on the length of the adjacent edge, length 1 corresponding to an infinitesimal rescaling). It turns out that R is a homotopy equivalence of operads and, since $W\mathcal{B}$ is homotopy equivalent to $\mathcal{B}$, this proves the proposition. □

In particular, the formality of the little balls operad will follow from that of the Fulton-MacPherson operad.

5.3. The canonical projections

Let V be a finite set containing A as a subset. Set $I = V \setminus A$. There is an obvious semi-algebraic map

$$\pi\colon \mathrm{C}[V] \longrightarrow \mathrm{C}[A] \tag{5.10}$$

given by forgetting from the configuration $y \in \mathrm{C}[V]$ all the points labeled by I. This map π can also be defined as an operad structure map. Indeed choose an arbitrary linear order on V and consider the inclusion $\iota\colon A \hookrightarrow V$ as a weak ordered partition. For $v \in V$, $\iota^{-1}(v)$ is either empty or a singleton $\{v\}$. Since $\mathrm{C}[\emptyset]$ and $\mathrm{C}[\{v\}]$ are both one-point spaces, the projection on the first factor

$$\mathrm{proj}\colon \mathrm{C}[V] \times \prod_{v\in V} \mathrm{C}[\iota^{-1}(v)] \xrightarrow{\cong} \mathrm{C}[V]$$

gives a homeomorphism which we use to identify these two spaces. Then the operad structure map

$$\mathrm{C}[V] = \mathrm{C}[V] \times \prod_{v\in V} \mathrm{C}[\iota^{-1}(v)] \xrightarrow{\Phi_\iota} \mathrm{C}[A]$$

is exactly the map π.

DEFINITION 5.3.1. The map $\pi\colon \mathrm{C}[V] \to \mathrm{C}[A]$ of (5.10) is called the *canonical projection (associated to the inclusion $A \subset V$).*

The Kontsevich configuration space integral will be defined through a pushforward of some minimal semi-algebraic forms along such canonical projections. For this to be possible, canonical projections have to be oriented SA bundles (that is, semi-algebraic bundles whose fibers are compact oriented manifolds; see Chapter 4 and [**18**, Definition 8.1]):

THEOREM 5.3.2. *Let A be a finite set and let I be a linearly ordered finite set disjoint from A. The canonical projection*

$$\pi\colon \mathrm{C}[A \cup I] \longrightarrow \mathrm{C}[A]$$

is an oriented SA bundle with fiber of dimension

$$\dim(\mathrm{fiber}(\pi)) \begin{cases} = N\cdot|I|, & \text{if } |A|\geq 2 \text{ or } I=\emptyset;\\ < N\cdot |I|, & \text{otherwise.}\end{cases}$$

Assume moreover that $|A| \geq 2$. Then the fiber of π is the space of configurations of $|I|$ points in $\mathbb{R}^N \setminus A$ compactified by adding a boundary to this open manifold.

- *When N is odd the orientation of the fiber of π depends on the linear order of I. A transposition of that linear order reverses the orientation.*
- *When N is even the orientation of the fiber is independent of the linear order on I.*

For example, when $|I| = 1$ and $|A| \geq 2$, the fiber of π is a closed N-ball with $|A|$ disjoint open balls removed from its interior.

The proof of this theorem is not very difficult but it is long. Since techniques used in the proof are not used anywhere else in the paper we decided to delay it until Section 5.9. Notice however that although $\mathrm{C}[n]$ are smooth manifolds with corners, it is not true that the canonical projections are smooth bundles, because their restrictions to the boundary are usually not submersions, as shown in Example 5.9.1. This is the reason why we have to work with semi-algebraic forms instead of smooth forms.

Canonical projections can also be used to construct retractions to the operad structure maps associated to a *non-degenerate* partition (see Definition 2.3.1) as in the following easy-to-prove proposition and corollary.

PROPOSITION 5.3.3. *Let $\nu\colon A \to P$ be an ordered weak partition and set $A_p = \nu^{-1}(p)$ for $p \in P$ as in the setting 2.4.1. For $q \in P$ denote by π_q the canonical projection associated to the inclusion $A_q \subset A$. Then the composition*

$$\mathrm{C}[P] \times \prod_{p\in P} \mathrm{C}[A_p] \xrightarrow{\Phi_\nu} \mathrm{C}[A] \xrightarrow{\pi_q} \mathrm{C}[A_q]$$

is the projection on that factor.

Suppose moreover that ν is non-degenerate, that is, it is surjective. Use any section of ν to identify P as a subset of A and let π_0 be the associated canonical projection. Then the composition

$$\mathrm{C}[P] \times \prod_{p\in P} \mathrm{C}[A_p] \xrightarrow{\Phi_\nu} \mathrm{C}[A] \xrightarrow{\pi_0} \mathrm{C}[P]$$

is the projection on the first factor.

COROLLARY 5.3.4. *If $\nu\colon A \to P$ is a non-degenerate ordered partition, then the operad structure map*

$$\Phi_\nu : \mathrm{C}[P] \times \prod_{p\in P} \mathrm{C}[A_p] \longrightarrow \mathrm{C}[A]$$

is injective and admits a continuous semi-algebraic retraction.

PROOF. A retraction is given by $(\pi_p)_{p\in P^*}$ where π_p is as in the previous proposition. □

This corollary is clearly wrong when the weak partition ν is degenerate.

5.4. Decomposition of the boundary of $\mathrm{C}[n]$ into codimension 0 faces

In this section we show that the boundary of $\mathrm{C}[n]$ decomposes as the union of the images of certain operad structure maps. Indeed, Proposition 5.4.1 below gives a partition of $\partial C[n]$ (up to codimension 1 intersections) whose pieces are images of "$\circ_i$" operations. Most of the operad structure on $\mathrm{C}[\bullet]$ can in fact be understood as an explicit decomposition of the boundary of $\mathrm{C}[n]$ as a union of faces homeomorphic to products of the form $\mathrm{C}[k] \times \mathrm{C}[n_1] \times \cdots \times \mathrm{C}[n_k]$. This is not true for the nullary part though.

Let V be a finite set. We will study the boundary of the manifold $\mathrm{C}[V]$. Recall that the elements of that boundary are characterized in Proposition 5.1.3. For a non-empty subset W of V, we will consider the configurations $y \in \mathrm{C}[V]$ such that the points $y(w)$ labeled by $w \in W$ are infinitesimally closer to each other with respect to any other point $y(v)$ labeled by $v \in V \setminus W$. We will show that these subsets of configurations give a decomposition of $\partial\,\mathrm{C}[V]$ into codimension 0 faces (Proposition 5.4.1) when W runs over proper subsets of cardinality ≥ 2.

For a non-empty subset $W \subset V$, let V/W be the quotient set of V in which all the elements of W are identified to a single element. In particular $|V/W| = |V| - |W| + 1$. Suppose given a linear order on V/W and consider the projection to the quotient

$$q\colon V \longrightarrow V/W$$

which is an ordered non-degenerate partition of V. One then has a structure map

$$\Phi_q\colon \mathrm{C}[V/W] \times \prod_{p\in V/W} \mathrm{C}[q^{-1}(p)] \longrightarrow \mathrm{C}[V].$$

Since $q^{-1}(p)$ is either a singleton $\{v\}$ or the subset W and since $\mathrm{C}[\{v\}]$ is a one-point space, we can identify the domain of Φ_q with $\mathrm{C}[V/W] \times \mathrm{C}[W]$. This defines a map

$$\Phi_W := \Phi_q\colon \mathrm{C}[V/W] \times \mathrm{C}[W] \longrightarrow \mathrm{C}[V] \tag{5.11}$$

that we will denote by Φ^V_W when we want to emphasize the set V.

In terms of operads, the map Φ_W corresponds to a "circle-i" operadic operation $\circ_i$, up to some permutation. Indeed, when $V = \{1, \dots, n+k\} = \underline{n+k}$ and $W = \{i, \dots, i+k\} \cong \underline{k+1}$ then $V/W \cong \underline{n}$ and Φ_W is exactly

$$\circ_i\colon \mathrm{C}[n] \times \mathrm{C}[k+1] \longrightarrow \mathrm{C}[n+k].$$

The image of Φ_W consists of configurations in $\mathrm{C}[V]$ such that the points labeled by W are infinitesimaly close to each other compared to any point labeled by $V \setminus W$. This condition is empty when $V = W$ or when W is a singleton; in other words for such a W the image of Φ_W is all of $\mathrm{C}[V]$. For proper subsets $W \subset V$ of cardinality ≥ 2, the image of Φ_W is in the boundary of $\mathrm{C}[V]$. Actually, the next proposition shows that the images of all these Φ_W supply a decomposition of $\partial\,\mathrm{C}[V]$. The pieces of this decomposition are indexed by the "boundary faces" set

$$\mathcal{BF}(V) := \{W \subset V : W \neq V \text{ and } |W| \geq 2\}. \tag{5.12}$$

PROPOSITION 5.4.1.

(i) The boundary of $\mathrm{C}[V]$ *decomposes as*

$$\partial\,\mathrm{C}[V] = \bigcup_{W \in \mathcal{BF}(V)} \mathrm{im}(\Phi_W);$$

(ii) For $W \in \mathcal{BF}(V)$,

$$\dim(\mathrm{im}(\Phi_W)) = N \cdot |V| - N - 2 = \dim(\partial\,\mathrm{C}[V]);$$

(iii) For $W_1 \neq W_2$ *in* $\mathcal{BF}(V)$,

$$\dim(\mathrm{im}(\Phi_{W_1}) \cap \mathrm{im}(\Phi_{W_2})) < N \cdot |V| - N - 2.$$

PROOF. (i) By Proposition 5.1.3, $\mathrm{im}(\Phi_W) \subset \partial\,\mathrm{C}[V]$ for $W \in \mathcal{BF}(V)$. We will prove that the boundary is contained in the union of the images of the Φ_W. Let $y \in \partial\,\mathrm{C}[V]$. By Proposition 5.1.3 there exist distinct elements $u_0, v_0, w_0 \in V$ such that

$$y(v_0) \simeq y(w_0) \,\mathrm{rel}\, y(u_0).$$

Set

$$W = \{w \in V : y(v_0) \simeq y(w) \,\mathrm{rel}\, y(u_0)\}.$$

Then $v_0, w_0 \in W$ and $u_0 \in V \setminus W$, and hence $W \in \mathcal{BF}(V)$. Consider the canonical projections

$$\pi_1\colon \mathrm{C}[V] \longrightarrow \mathrm{C}[(V \setminus W) \cup \{w_0\}] \cong \mathrm{C}[V/W] \quad \text{and} \quad \pi_2\colon \mathrm{C}[V] \longrightarrow \mathrm{C}[W].$$

Then $y = \Phi_W(\pi_1(y), \pi_2(y))$. This proves (i).

(ii) For $W \in \mathcal{BF}(V)$, the map Φ_W is injective (by Corollary 5.3.4) and hence, by compactness, it is a homeomorphism onto its image. Since $|W| \geq 2$ and $|V/W| \geq 2$, Proposition 5.1.2 implies that

$$\begin{aligned}\dim(\mathrm{im}\,\Phi_W) &= \dim \mathrm{C}[V/W] + \dim \mathrm{C}[W] \\ &= (N \cdot |V/W| - N - 1) + (N \cdot |W| - N - 1 \\ &= N \cdot |V| - N - 2.\end{aligned}$$

(iii) Let $W_1, W_2 \in \mathcal{BF}(V)$ with $W_1 \neq W_2$. We consider three cases.

Case 1: Suppose that $W_1 \cap W_2 = \emptyset$. Then $\mathrm{im}(\Phi_{W_1}) \cap \mathrm{im}(\Phi_{W_2})$ is the image of the composition

$$\mathrm{C}[(V/W_2)/W_1] \times \mathrm{C}[W_1] \times \mathrm{C}[W_2] \xrightarrow{\left(\Phi_{W_1}^{V/W_2}\right)\times \mathrm{id}} \mathrm{C}[V/W_2] \times \mathrm{C}[W_2] \xrightarrow{\Phi_{W_2}^{V}} \mathrm{C}[V]$$

and an analogous computation as in (ii) implies that this image is of dimension $N \cdot |V| - N - 3$.

Case 2: Suppose that $W_1 \subset W_2$ (or the other way around). Then $\mathrm{im}(\Phi_{W_1}) \cap \mathrm{im}(\Phi_{W_2})$ is the image of the composition

$$\mathrm{C}[V/W_2] \times \mathrm{C}[W_2/W_1] \times \mathrm{C}[W_1] \xrightarrow{\mathrm{id}\times\left(\Phi_{W_1}^{W_2}\right)} \mathrm{C}[V/W_2] \times \mathrm{C}[W_2] \xrightarrow{\Phi_{W_2}^{V}} \mathrm{C}[V]$$

and again this image is of dimension $N \cdot |V| - N - 3$.

Case 3: Suppose that $W_1 \cap W_2 \neq \emptyset$, $W_1 \not\subset W_2$, and $W_2 \not\subset W_1$. Choose $a \in W_1 \cap W_2$, $b \in W_1 \setminus W_2$, and $c \in W_2 \setminus W_1$. For $y \in \mathrm{im}(\Phi_{W_1}) \cap \mathrm{im}(\Phi_{W_2})$ we simultaneously have

$$y(a) \simeq y(b) \,\mathrm{rel}\, y(c) \quad \text{and} \quad y(a) \simeq y(c) \,\mathrm{rel}\, y(b),$$

which is impossible. Thus $\mathrm{im}(\Phi_{W_1}) \cap \mathrm{im}(\Phi_{W_2})$ is empty.

□

More generally the operad structure maps

$$\Phi_\nu \colon \mathrm{C}[k] \times \mathrm{C}[n_1] \times \cdots \times \mathrm{C}[n_k] \longrightarrow \mathrm{C}[n]$$

map homeomorphically to faces of codimension $(k-2)$ in the boundary $\partial\,\mathrm{C}[n]$ when $2 \leq k < n$, $n = n_1 + \cdots + n_k$, and $n_1, \ldots, n_k \geq 1$. This in fact gives a complete stratification of that boundary, but we will not use this fact. However, when $n_i = 0$ for some $1 \leq i \leq k$, then Φ_ν is not an inclusion, and in this case the study of Φ_ν can require a more careful treatment as will be the case for example in Section 5.6.

5.5. Spaces of singular configurations

REMARK 5.5.1. This and the next four sections discuss some of the more technical properties of the Fulton-MacPherson operad which will be needed for the corresponding technical parts of the proof of the properties of the Kontsevich configuration space integral in Chapter 9. The reader can thus safely skip Sections 5.5-5.9 for the time being and jump to Chapter 6, except for the notion of *condensation* in Definition 5.6.1 which is necessary for defining the cooperad structure on the space of diagrams in Chapter 7.

At times we will need to consider variations of the configuration spaces $\mathrm{C}[V]$ in which some components of a configuration are allowed to coincide exactly, that is, without extra infinitesimal information to distinguish the points. The goal of this section is to make this situation precise.

Let A, I_1, I_2 be disjoint finite sets. Set $V_i = A \cup I_i$ for $i = 1, 2$ and $V = A \cup I_1 \cup I_2$. Hence we have a pushout of sets $V = V_1 \cup_A V_2$. Consider the following pullback where π_1 and π_2 are canonical projections:

$$\begin{array}{ccc} \mathrm{C}^{\mathrm{sing}}[V_1, V_2] & \xrightarrow{q_1} & \mathrm{C}[V_1] \\ {\scriptstyle q_2}\big\downarrow & \text{pullback} & \big\downarrow{\scriptstyle \pi_1} \\ \mathrm{C}[V_2] & \xrightarrow[\pi_2]{} & \mathrm{C}[A]. \end{array} \tag{5.13}$$

Intuitively, $\mathrm{C}^{\mathrm{sing}}[V_1, V_2]$ can be seen as a compactified singular space of configurations of points in $\mathbb{R}^N$ labeled by $v \in V$. By "singular" we mean that, for a configuration y, the component $y(i_1)$ labeled by $i_1 \in I_1$ may coincide with another component $y(i_2)$ labeled by $i_2 \in I_2$.

Since $V_i \subset V$, we have for $i = 1, 2$ the canonical projections

$$\rho_i \colon \mathrm{C}[V] \longrightarrow \mathrm{C}[V_i].$$

As $\pi_1\rho_1 = \pi_2\rho_2$, we have a surjective map

$$\rho \colon \mathrm{C}[V] \longrightarrow \mathrm{C}^{\mathrm{sing}}[V_1, V_2] \tag{5.14}$$

to the pullback induced by (ρ_1, ρ_2). Intuitively, when $y(i_1)$ and $y(i_2)$ are infinitesimally close in $y \in \mathrm{C}[V]$, $\rho(y)$ is the singular configuration in which we forget the infinitesimal data associated to those components.

Consider the canonical projections $\pi \colon \mathrm{C}[V] \to \mathrm{C}[A]$ and $\pi_{V_1} \colon \mathrm{C}[V] \to \mathrm{C}[V_1]$, and the composition

$$\pi' := q_1 \circ \pi_1 = q_2 \circ \pi_2 \colon \mathrm{C}^{\mathrm{sing}}[V_1, V_2] \longrightarrow \mathrm{C}[A].$$

Recall the notation $[\![M]\!]$ for semi-algebraic chains from (4.1), (4.2), and (4.7) in Chapter 4.

LEMMA 5.5.2. *There is a commutative diagram*

$$\begin{array}{ccc} \mathrm{C}[V] & \xrightarrow{\rho} & \mathrm{C}^{\mathrm{sing}}[V_1, V_2] \\ & {\scriptstyle \pi_{V_1}}\searrow \quad \swarrow{\scriptstyle q_1} & \\ & \mathrm{C}[V_1] & \end{array}$$

where π_{V_1} and q_1 are orientable SA bundles. If moreover $|V_1| \geq 2$, then for each $x \in \mathrm{C}[V_1]$

$$\rho_*\left([\![{\pi_{V_1}}^{-1}(x)]\!]\right) = \pm[\![{q_1}^{-1}(x)]\!].$$

In other words, ρ induces a map of degree ± 1 between the fibers of π_{V_1} and q_1.

Similarly there is a commutative diagram

$$\begin{array}{ccc} \mathrm{C}[V] & \xrightarrow{\rho} & \mathrm{C}^{\mathrm{sing}}[V_1, V_2] \\ & {\scriptstyle \pi}\searrow \quad \swarrow{\scriptstyle \pi'} & \\ & \mathrm{C}[A], & \end{array}$$

and if $|A| \geq 2$, then ρ induces a map of degree ± 1 between the fibers of π and π'.

PROOF. Theorem 5.3.2 states that canonical projections are oriented SA bundles, and hence so are π_{V_1} and π_2. Therefore q_1 is also an oriented SA bundle as the pullback of π_2 along π_1 [**18**, Proposition 8.4]. When $|V_1| \geq 2$, the fiber ${\pi_{V_1}}^{-1}(x)$ of

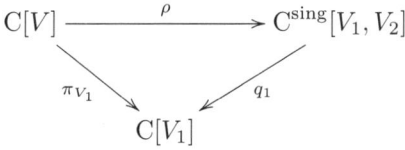

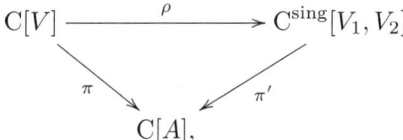

π_{V_1} over any $x \in \mathrm{C}[V_1]$ is a compact manifold whose interior can be identified with the space of injections

$$\mathrm{Inj}(I_2, \mathbb{R}^N \setminus V_1) = \{y \colon I_2 \hookrightarrow \mathbb{R}^N \setminus V_1\}$$

where V_1 is seen as a fixed subset in $\mathbb{R}^N$. From the pullback (5.13) the fiber of q_1 is the same as the fiber of π_2 whose interior can similarly be identified with

$$\mathrm{Inj}(I_2, \mathbb{R}^N \setminus A).$$

Thus ρ maps the interior of the fiber ${\pi_{V_1}}^{-1}(x)$ homeomorphically to a dense subset of the fiber ${q_1}^{-1}(x)$, and hence induces a degree ± 1 map between the fibers of π_{V_1} and q_1.

The proof of the second part of the lemma is similar. □

5.6. Pullback of a canonical projection along an operad structure map

In Chapter 9 we will define the Kontsevich configuration space integral I along the lines of (1.8) in the Introduction, and will want to prove that it is a morphism of (almost) cooperads. Since this integral is defined using pushforward along a canonical projection, we need to investigate the pullback of a canonical projection along an operad structure map, as in Diagram (5.15) below. This is the aim of this section. The main results are Proposition 5.6.2 (complemented by Proposition 5.6.6) and Proposition 5.6.5. This section is technical and is only needed in Section 9.5, except for the notion of *condensation* in Definition 5.6.1, which, as mentioned before, is needed to define the cooperad stucture on the space of diagrams.

Throughout this section we fix a weak ordered partition $\nu \colon A \to P$ and set

$$P^* = \{0\} \olessthan P, \quad A_p = \nu^{-1}(p), \quad \text{and } A_0 = P$$

as in the setting 2.4.1. We also have an associated operad structure map

$$\Phi_\nu \colon \mathrm{C}[P] \times \prod_{p \in P} \mathrm{C}[A_p] = \prod_{p \in P^*} \mathrm{C}[A_p] \longrightarrow \mathrm{C}[A]$$

from (5.9). We also fix a linearly ordered finite set I disjoint from A and P and set $V = A \cup I$. Thus we can consider the canonical projection

$$\pi \colon \mathrm{C}[V] \longrightarrow \mathrm{C}[A]$$

associated to $A \subset V$ as in (5.10). The elements of $I := V \setminus A$ will be called *internal vertices*, the elements of A *external vertices*, and the elements of V *vertices*. As the case $|A| \leq 1$ is somewhat degenerate and has to be treated separately, we will always in this section assume that $|A| \geq 2$.

Define $\mathrm{C}[V, \nu]$ as the pullback

$$\begin{array}{ccc} \mathrm{C}[V,\nu] & \xrightarrow{\ \Phi'_\nu\ } & \mathrm{C}[V] \\ {\scriptstyle \pi'_\nu}\downarrow & \text{pullback} & \downarrow{\scriptstyle \pi} \\ \prod_{p \in P^*} \mathrm{C}[A_p] & \xrightarrow[\ \Phi_\nu\ ]{} & \mathrm{C}[A], \end{array} \tag{5.15}$$

where π is the canonical projection (5.10) and Φ_ν is the operad structure map (5.9).

The main goal of this section is to show that this pullback decomposes as a union

$$\mathrm{C}[V,\nu] = \bigcup_{\lambda} \mathrm{C}[V,\lambda] \tag{5.16}$$

(Proposition 5.6.2) such that the restrictions $\Phi'_\lambda := \Phi'_\nu | \mathrm{C}[V,\lambda]$ are closely related to some operad structure maps Φ'_λ (Proposition 5.6.5). Moreover (5.16) is "almost" a partition, in the sense that the intersections $\mathrm{C}[V,\lambda] \cap \mathrm{C}[V,\mu]$ are of lower dimension for $\lambda \neq \mu$ (Proposition 5.6.6).

Let us first give a rough idea of how we will show this. To make it easier, let us temporarily make an additional assumption that ν is non-degenerate (that is, each A_p is non-empty) and that P contains at least two elements. In that case, the map Φ_ν is the inclusion of some part of the boundary of $\mathrm{C}[A]$. More precisely, $\mathrm{im}(\Phi_\nu)$ consists of all configurations $x \in \mathrm{C}[A]$ such that, for $a, b, c \in A$, if $\nu(a) = \nu(b) \neq \nu(c)$ then $x(a) \simeq x(b) \,\mathrm{rel}\, x(c)$. We will say that such a configuration $x \in \mathrm{C}[A]$ is *ν-condensed.* In other words, a configuration $x \in \mathrm{im}(\Phi_\nu)$ can be thought of as a family indexed by $p \in P$ of clusters of points, where the p-th cluster consist of points $x(a)$ indexed by $a \in A_p = \nu^{-1}(p)$. For example, the configuration $x \in \mathrm{C}[6]$ from Figure 5.2 in Section 5.2 is ν-condensed for the partition ν given at beginning of Example 5.2.1.

As Φ_ν is an inclusion (because of our extra assumption), the pullback $\mathrm{C}[V,\nu]$ is the subset of $\mathrm{C}[V]$ consisting of all configurations $y \in \mathrm{C}[V]$ such that $x := \pi(y)$ is ν-condensed. Consider such a $y \in \mathrm{C}[V,\nu]$. One can then look at the position of the points $y(i)$, for $i \in I$, with respect to the various clusters of points $\{x(a) : a \in A_p\}$, for $p \in P$. Such a point $y(i)$ could be inside or infinitesimally close to some cluster indexed by $p \in P$, in which case we say that, for this configuration, i is *p-local*; or $y(i)$ could be close to none of the clusters in which case we say that i is *global.* These cases can be encoded by a function

$$\lambda\colon I \longrightarrow P^*$$

with $\lambda(i) = p$ if i is p-local, and $\lambda(i) = 0$ if i is global. It is natural to extend λ to V by letting $\lambda|A = \nu$. Such a map $\lambda\colon V \to P^*$ will be called a *condensation* (Definition 5.6.1 below), and there is a natural partition of $\mathrm{C}[V,\nu]$ as a union of the subspaces $\mathrm{C}[V,\lambda]$ consisting of λ-condensed configurations y. Moreover, under our extra assumption, each $\mathrm{C}[V,\lambda]$ is homeomorphic to the product $\prod_{p\in P^*} \mathrm{C}[V_p]$ where $V_0 = \lambda^{-1}(0) \cup P$ and $V_p = \lambda^{-1}(p)$ for $p \in P$, and through this homeomorphism the restriction $\Phi'_\nu | \mathrm{C}[V,\lambda]$ is an operad structure map.

The precise description of the decomposition of $\mathrm{C}[V,\nu]$ is a bit more delicate when the weak partition ν is degenerate, that is, when our extra assumption does not hold. We now proceed with the details and first define the notion of a condensation.

Definition 5.6.1. Let A be a finite set, $\nu\colon A \to P$ be a weak ordered partition, I be a finite linearly ordered set disjoint from A, $P^* := \{0\} \oslash P$, and $V := A \amalg I$. Set $A_p = \nu^{-1}(p)$ for $p \in P$ and $A_0 = P$. Elements of V are called *vertices* as above.

- A *condensation of V relative to ν* is a map

$$\lambda\colon V \longrightarrow P^*$$

such that $\lambda|A = \nu$.

- The set of all such condensations λ is denoted by $\mathrm{Cond}(V,\nu)$, or simply $\mathrm{Cond}(V)$ when ν is understood.
- Given a condensation $\lambda \in \mathrm{Cond}(V)$, a vertex $v \in V$ is *p-local* if $\lambda(v) = p$ for some $p \in P$, and it is *global* if $\lambda(v) = 0$.
- A configuration $y \in \mathrm{C}[V]$ is *λ-condensed* if for each $u, v, w \in V$ and $p \in P$ such that u and v are p-local and w is not p-local we have $y(u) \simeq y(v) \operatorname{rel} y(w)$.
- A condensation $\lambda \in \mathrm{Cond}(V,\nu)$ is *essential* if for each $p \in \lambda(I)$ we have that $|A_p| \geq 2$. We denote the set of essential condensations by $\mathrm{EssCond}(V,\nu)$, or simply $\mathrm{EssCond}(V)$ when ν is understood.

The terminology *condensation* comes from the idea that a λ-condensed configuration $x \in \mathrm{C}[V]$ consists of clusters of points condensed together according of the values of λ on their vertices.

It is easy to convince oneself that a configuration $y \in \mathrm{C}[V]$ is λ-condensed if and only if it is in the image of an operad structure map $\Phi_{\widehat{\lambda}}$, where $\widehat{\lambda}$ is some weak partition of V constructed from λ (see (5.22) and (5.23) below for definitions of $\widehat{\lambda}$ and $\Phi_{\widehat{\lambda}}$).

A condensation is essential if there are no internal p-vertices when $|A_p| \leq 1$, $p \in P$, and no global (internal) vertices when $|P| \leq 1$. We will see latter that non-essential condensations are in some sense negligible. For example they are not needed in the decomposition of $\mathrm{C}[V,\nu]$ in Proposition 5.6.2 below and their contribution to the Kontsevich configuration space integral is zero as we will see in Lemma 9.5.3.

For λ a condensation of V relative to ν, set

$$\mathrm{C}[V,\lambda] := \{g \in \mathrm{C}[V,\nu] : \Phi'_\nu(g) \text{ is } \lambda\text{-condensed}\} \tag{5.17}$$

where $\mathrm{C}[V,\nu]$ and Φ'_ν are from (5.15).

Recall that in this section we assume $|A| \geq 2$. Our first important result is the following decomposition of the pullback $\mathrm{C}[V,\nu]$.

PROPOSITION 5.6.2. *For $|A| \geq 2$, there is a decomposition*

$$\mathrm{C}[V,\nu] = \bigcup_{\lambda \in \mathrm{EssCond}(V,\nu)} \mathrm{C}[V,\lambda]$$

where λ runs over all essential condensations relative to ν.

PROOF. Recall the pullback $\mathrm{C}[V,\nu]$ of diagram (5.15) and let

$$g = (y, (x_p)_{p \in P^*}) \in \mathrm{C}[V,\nu]$$

with $y \in \mathrm{C}[V]$, $x_p \in \mathrm{C}[A_p]$, and $\pi(y) = \Phi_\nu\left((x_p)_{p\in P^*}\right)$. We need to construct an essential condensation λ such that $g \in \mathrm{C}[V,\lambda]$. For $i \in I$ and $p \in P$ we say that i is *p-local for g* if

(i) $|A_p| \geq 2$, and
(ii) $\forall a, b \in A : (\nu(a) = p \text{ and } \nu(b) \neq p) \Longrightarrow (y(a) \simeq y(i) \operatorname{rel} y(b))$.

If i is p-local, then it cannot be q-local for $q \neq p$ because otherwise there would exist $a \in A_p$ and $b \in A_q$ (since $|A_p|, |A_q| \geq 2$), with both $y(a) \simeq y(i) \operatorname{rel} y(b)$ and $y(b) \simeq y(i) \operatorname{rel} y(a)$, which is impossible.

Define a condensation $\lambda\colon V \to P^*$ by

$$\lambda(v) = \begin{cases} \nu(v), & \text{if } v \in A; \\ p, & \text{if } v \in I \text{ and } v \text{ is } p\text{-local for } g, \text{ for some } p \in P; \\ 0, & \text{if } v \in I \text{ and there is no } p \in P \text{ for which } v \text{ is } p\text{-local for } g. \end{cases}$$

Let us show that λ is essential. If $p \in \lambda(I) \cap P$ then $|A_p| \geq 2$ by condition (i). If $0 \in \lambda(I)$ then $|P| \geq 2$ because otherwise P is a singleton $\{p_1\}$ (as P cannot be empty since $|A| \geq 2$), in which case every $i \in I$ is p_1-local (except if $|A_{p_1}| < 2$ which is again impossible since we assume in this section that $|A| \geq 2$). Therefore λ is an essential condensation relative to ν.

We now show that $g \in \mathrm{C}[V, \lambda]$. Let $u, v, w \in V$ such that $\lambda(u) = \lambda(v) = p \neq 0$ and $\lambda(w) \neq p$. We need to show that

$$y(u) \simeq y(v) \operatorname{rel} y(w). \tag{5.18}$$

If $|A_p| \leq 1$ then $u, v \in A_p$ since no internal vertex can be p-local due to (i). Hence $u = v$ and (5.18) is obvious. Suppose now that $|A_p| \geq 2$. As $\lambda(w) \neq p$, there are two cases:

(A) $w \in A_q$ for some $q \neq p$, or
(B) $w \in I$ and there exists $a \in A_p$ and $b \in A_q$ for some $q \neq p$ such that

$$y(a) \not\simeq y(w) \operatorname{rel} y(b).$$

In case (A) we can pick $a \in A_p$ and, whether $u \in I$ or $u \in A_p$, we have $y(a) \simeq y(u) \operatorname{rel} y(w)$. Similarly $y(a) \simeq y(v) \operatorname{rel} y(w)$. By transitivity we get (5.18).

In case (B) we have $y(a) \not\simeq y(w) \operatorname{rel} y(b)$. Since $y(a) \simeq y(u) \operatorname{rel} y(b)$ we deduce that $y(a) \simeq y(u) \operatorname{rel} y(w)$. Similarly $y(a) \simeq y(v) \operatorname{rel} y(w)$. Again by transitivity we get (5.18). □

The various $\mathrm{C}[V, \lambda]$ that appear in the union of the previous proposition are not necessarily pairwise disjoint. However, their intersection is of positive codimension as we will see in Proposition 5.6.6. We have assumed in this section that $|A| \geq 2$; when $|A| < 2$ it is possible that there are no essential condensations at all, in which case the decomposition from the last proposition cannot hold.

Let $\lambda\colon V \to P^* = P \cup \{0\}$ be an essential condensation. Our next goal is to show that the restriction of $\Phi'_\nu\colon \mathrm{C}[V, \nu] \to \mathrm{C}[V]$ to $\mathrm{C}[V, \lambda]$ is closely related to a map

$$\Phi'_\lambda\colon \prod_{p \in P^*} \mathrm{C}[V_p] \longrightarrow \mathrm{C}[V],$$

where $V_p = \lambda^{-1}(p)$ for $p \in P$, $V_0 = \lambda^{-1}(0) \cup P$, and the map Φ'_λ can be identified with an explicit operad structure map $\Phi_{\widehat{\lambda}}$. Here $\widehat{\lambda}$ is some refined partition of λ. In short, this amounts to saying that λ-condensed configurations are exactly the image of a certain operad structure map. This is the content of Proposition 5.6.5.

To prove this, we will need to construct various maps that are collected in the following diagram for reader's convenience, along with numbers of equations where they can be found. The two identifications are due to the fact that $\prod_{i \in I_0} \mathrm{C}[\{i\}]$ is a one-point space.

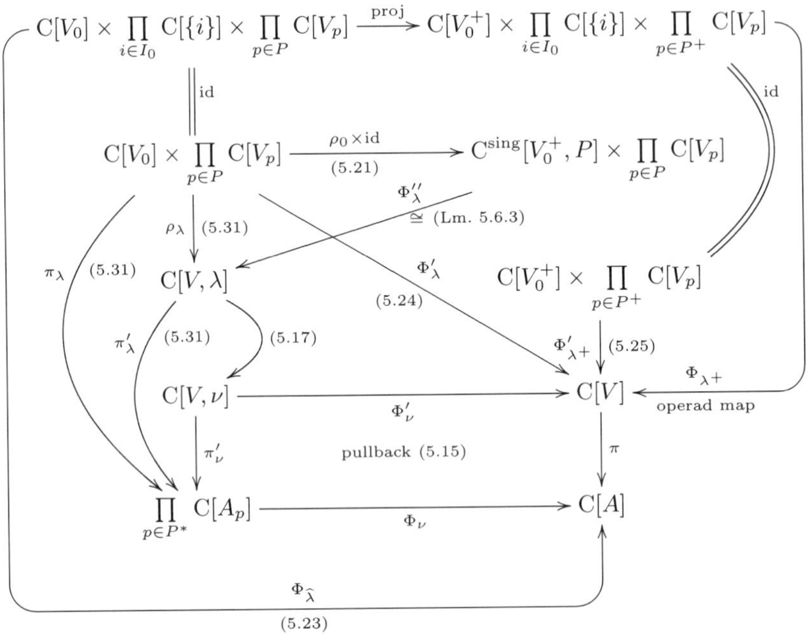

First we show that $C[V, \lambda]$ is homeomorphic to the product of configuration spaces $C[V_p]$, $p \in P$, and another, maybe singular, configuration space $C^{\mathrm{sing}}[V_0^+, P]$. Let us construct the V_p's. Recall that $A_p = \nu^{-1}(p) = A \cap \lambda^{-1}(p)$ for $p \in P$, $A_0 = P$, and $V = A \amalg I$. For $p \in P^*$, set

(5.19) $$I_p = I \cap \lambda^{-1}(p) \quad \text{and} \quad V_p = A_p \cup I_p,$$

so $V_p = \lambda^{-1}(p)$ for $p \in P$, and $V_0 = \lambda^{-1}(0) \cup P$. The linear order of I restricts to linear orders on I_p for $p \in P^*$. Moreover we order V_0 as (remember that $A_0 = P$)

$$V_0 = I_0 \otimes P.$$

Also define the subsets

$$P^+ := \{p \in P : A_p \neq \emptyset\} \subset P$$

and

$$V_0^+ := I_0 \cup P^+ \subset V_0.$$

Hence we have a pushout of sets $V_0 = V_0^+ \cup_{P^+} P$. Consider the following pullback

(5.20) $$\begin{array}{ccc} C^{\mathrm{sing}}[V_0^+, P] & \xrightarrow{q_{V_0^+}} & C[V_0^+] \\ {\scriptstyle q_P}\downarrow & \text{pullback} & \downarrow{\scriptstyle \pi^+} \\ C[P] & \xrightarrow{\pi'^+} & C[P^+], \end{array}$$

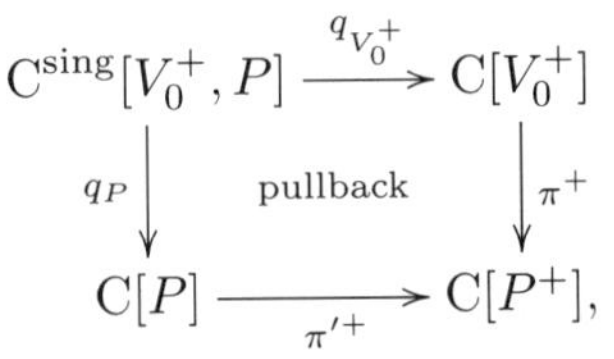

where π^+ and π'^+ are the canonical projections. This defines a singular configuration space as in Section 5.5. When ν is non-degenerate then $P^+ = P$, $V_0^+ = V_0$, and

$\mathrm{C}^{\mathrm{sing}}[V_0^+, P]$ is just the configuration space $\mathrm{C}[V_0]$. In any case we have an induced map, as in (5.14),

$$\rho_0\colon \mathrm{C}[V_0] \longrightarrow \mathrm{C}^{\mathrm{sing}}[V_0^+, P], \tag{5.21}$$

which, by Lemma 5.5.2, induces a degree ± 1 map between the fibers.

Define the weak ordered partition

$$\widehat{\lambda}\colon V \longrightarrow V_0 \tag{5.22}$$

$$v \longmapsto \widehat{\lambda}(v) = \begin{cases} v, & \text{if } \lambda(v) = 0; \\ \lambda(v), & \text{otherwise.} \end{cases}$$

There is an associated operad structure map

$$\Phi_{\widehat{\lambda}}\colon \mathrm{C}[V_0] \times \left(\prod_{i \in I_0} \mathrm{C}[\{i\}] \times \prod_{p \in P} \mathrm{C}[V_p] \right) \longrightarrow \mathrm{C}[V]. \tag{5.23}$$

Since $\mathrm{C}[\{i\}]$ are one-point spaces, the domain of $\Phi_{\widehat{\lambda}}$ is homeomorphic to $\prod_{p \in P^*} \mathrm{C}[V_p]$ (through the obvious projection), and the composition of this homeomorphism with $\Phi_{\widehat{\lambda}}$ gives a map

$$\Phi'_\lambda\colon \prod_{p \in P^*} \mathrm{C}[V_p] \longrightarrow \mathrm{C}[V]. \tag{5.24}$$

We next show that Φ'_λ factors through the composition of $\Phi'_\nu | \mathrm{C}[V, \lambda]$ with a homeomorphism Φ''_λ between $\mathrm{C}^{\mathrm{sing}}[V_0^+, P] \times \prod_{p \in P} \mathrm{C}[V_p]$ and $\mathrm{C}[V, \lambda]$.

By definition $P^+ = \mathrm{im}(\nu)$ and, since λ is essential, $\mathrm{im}(\lambda) \subset P^+ \cup \{0\}$. Therefore the weak partition $\widehat{\lambda}$ factors as the composition of an ordered non-degenerate partition

$$\lambda^+ \colon V \longrightarrow V_0^+$$

and the inclusion $V_0^+ \hookrightarrow V_0$.

For $p \in P \setminus P^+$ we have $A_p = \emptyset$, and hence $V_p = \emptyset$ because λ is essential, and so $\mathrm{C}[V_p] = *$. Also $\mathrm{C}[\{i\}] = *$ for $i \in I_0$. Thus the projections induce a homeomorphism

$$\mathrm{C}[V_0^+] \times \prod_{p \in P} \mathrm{C}[V_p] \cong \mathrm{C}[V_0^+] \times \prod_{i \in I_0} \mathrm{C}[\{i\}] \times \prod_{p \in P^+} \mathrm{C}[V_p].$$

The composition of this homeomorphism with the operad structure map Φ_{λ^+} is a map

$$\Phi'_{\lambda^+}\colon \mathrm{C}[V_0^+] \times \prod_{p \in P} \mathrm{C}[V_p] \longrightarrow \mathrm{C}[V]. \tag{5.25}$$

Recall $q_{V_0^+}$ and q_P from (5.20), let

$$\pi_p\colon \mathrm{C}[V_p] \to \mathrm{C}[A_p]$$

be the canonical projections, and consider the two maps

$$\Phi'_{\lambda^+} \circ (q_{V_0^+} \times \mathrm{id})\colon \mathrm{C}^{\mathrm{sing}}[V_0^+, P] \times \prod_{p \in P} \mathrm{C}[V_p] \longrightarrow \mathrm{C}[V]$$

and

$$q_P \times (\times_{p \in P} \pi_p)\colon \mathrm{C}^{\mathrm{sing}}[V_0^+, P] \times \prod_{p \in P} \mathrm{C}[V_p] \longrightarrow \mathrm{C}[P] \times \prod_{p \in P} \mathrm{C}[A_p] = \prod_{p \in P^*} \mathrm{C}[A_p].$$

They induce a map

$$\Phi''_\lambda\colon \mathrm{C}^{\mathrm{sing}}[V_0^+, P] \times \prod_{p\in P} \mathrm{C}[V_p] \longrightarrow \mathrm{C}[V,\nu] \tag{5.26}$$

into the pullback (5.15).

LEMMA 5.6.3. Φ''_λ *of* (5.24) *is a homeomorphism onto* $\mathrm{C}[V,\lambda] \subset \mathrm{C}[V,\nu]$.

PROOF. Since the domain of Φ''_λ is compact, it is enough to prove that Φ''_λ is injective and that its image is $\mathrm{C}[V,\lambda]$. This is in fact not hard to see using the pictorial interpretations of virtual configurations. Here is a more formal proof.

For injectivity, let

$$z = (z_0, (z_p)_{p\in P}) \text{ and } z' = (z'_0, (z'_p)_{p\in P}) \in \mathrm{C}^{\mathrm{sing}}[V_0^+, P] \times \prod_{p\in P} \mathrm{C}[V_p]$$

be such that $\Phi''_\lambda(z) = \Phi''_\lambda(z')$. Since $\pi'_\nu \circ \Phi''_\lambda = q_P \times \left(\times_{p\in P} \pi_p\right)$, we get that

$$q_P(z_0) = q_P(z'_0). \tag{5.27}$$

As λ^+ is a non-degenerate partition, by Corollary 5.3.4 Φ'_{λ^+} is injective. Since $\Phi'_\nu \circ \Phi''_\lambda = \Phi'_{\lambda^+} \circ (q_{V_0^+} \times \mathrm{id})$, we deduce that

$$(q_{V_0^+}(z_0), (z_p)_{p\in P}) = (q_{V_0^+}(z'_0), (z'_p)_{p\in P}). \tag{5.28}$$

From (5.27) and (5.28) we deduce that $z = z'$ using the pullback diagram (5.20).

The image of $\Phi_{\hat\lambda}$ consists of λ-condensed configurations, and hence $\mathrm{im}(\Phi''_\lambda) \subset \mathrm{C}[V,\lambda]$. Let us prove surjectivity. Let

$$g = (y, (x_p)_{p\in P^*}) \in \mathrm{C}[V,\lambda].$$

As $y \in \mathrm{C}[V]$ is λ-local it belongs in the image of Φ'_λ. Since

$$\Phi'_\lambda = \Phi'_{\lambda^+} \circ (q_{V_0^+} \times \mathrm{id}) \circ (\rho_0 \times \mathrm{id}),$$

we can set

$$y = \Phi'_{\lambda^+}(z_0^+, (z_p)_{p\in P}) \text{ for some } (z_0^+, (z_p)_{p\in P}) \in \mathrm{C}[V_0^+] \times \prod_{p\in P} \mathrm{C}[V_p].$$

Since $\pi(y) = \Phi_\nu((x_p)_{p\in P^*})$, using π^+ and π'^+ from (5.20), we deduce that $\pi^+(z_0^+) = \pi'^+(x_0)$ (this can be seen for example by factoring ν through a non-degenerate partition $\nu^+\colon A \to P^+$ and using Corollary 5.3.4). Set $z_0 = (z_0^+, x_0) \in \mathrm{C}^{\mathrm{sing}}[V_0^+, P]$. Then $g = \Phi''_\lambda(z_0, (z_p)_{p\in P})$. □

Define the product of canonical projections

$$\pi_\lambda := \times_{p\in P^*} \pi_p\colon \prod_{p\in P^*} \mathrm{C}[V_p] \longrightarrow \prod_{p\in P^*} \mathrm{C}[A_p] \tag{5.29}$$

and the restriction

$$\pi'_\lambda := (\pi'_\nu | \mathrm{C}[V,\lambda])\colon \mathrm{C}[V,\lambda] \longrightarrow \prod_{p\in P^*} \mathrm{C}[A_p] \tag{5.30}$$

where π'_ν is from (5.15). Define also

$$\rho_\lambda := \Phi''_\lambda \circ (\rho_0 \times \mathrm{id})\colon \prod_{p\in P^*} \mathrm{C}[V_p] \longrightarrow \mathrm{C}[V,\lambda].$$

So we get the following commutative diagram we are aiming for

$$\begin{array}{ccc} \prod_{p\in P^*} \mathrm{C}[V_p] & \xrightarrow{\quad\rho_\lambda\quad} & \mathrm{C}[V,\lambda] \\ & {\scriptstyle\pi_\lambda}\searrow \quad \swarrow{\scriptstyle\pi'_\lambda} & \\ & \prod_{p\in P^*} \mathrm{C}[A_p]. & \end{array} \tag{5.31}$$

LEMMA 5.6.4. *π_λ and π'_λ are orientable SA bundles with fibers of dimension $N \cdot |I|$.*

PROOF. For π_λ, this is a direct consequence of Theorem 5.3.2 since π_λ is a product of canonical projections. Since λ is essential, $|A_p| \geq 2$ or $I_p = \emptyset$ for each $p \in P^*$, which yields the formula for the dimension of the fiber.

For π'_λ, the result comes from the homeomorphism Φ''_λ (Lemma 5.6.3) through which π'_λ can be identified with $q_P \times \underset{p\in P}{\times} \pi_p$, and from the fact that q_P is also an oriented SA bundle as it is the pullback of the canonical projection π^+ along π'^+ in Diagram (5.20). $\square$

We fix the orientations of the fibers of π_λ and π'_λ as follows. For the fibers of π_λ, we orient them as the product, in the linear order of P^*, of the fibers of π_p oriented as in Theorem 5.3.2 with respect to the linear order of I_p restricted from that of I. For the fibers of π'_λ, they are connected codimension 0 submanifolds of the fibers of π'_ν, which are canonically identified with the fibers of π because of the pullback (5.15). We orient then the fibers of π'_λ by the orientation of the fibers of π defined in Theorem 5.3.2 from the given linear order on I.

We will see that ρ_λ in (5.31) induces a change of orientation of the fibers according to the sign

$$\sigma(I,\lambda) := (-1)^{N\cdot|S(I,\lambda)|} \tag{5.32}$$

where

$$S(I,\lambda) := \{(v,w) \in I \times I : v < w \text{ and } \lambda(v) > \lambda(w)\}. \tag{5.33}$$

Recall the fundamental class of the fiber of an oriented SA bundle as in (4.7). The second main result of this section can then be summarized in the following

PROPOSITION 5.6.5. *Let $\lambda \in \mathrm{EssCond}(V,\nu)$ be an essential condensation relative to ν and consider Diagram (5.31) above.*

(i) *π_λ and π'_λ are oriented SA bundles with fibers of dimension $N \cdot |I|$.*
(ii) *ρ_λ induces a map of degree $\sigma(I,\lambda) = \pm 1$ between the fibers. More precisely, for $x \in \prod_{p\in P^*} \mathrm{C}[A_p]$,*

$$\rho_{\lambda *}([\![\pi_\lambda^{-1}(x)]\!]) = \sigma(I,\lambda) \cdot [\![{\pi'_\lambda}^{-1}(x)]\!]$$

in $\mathrm{C}_*(\pi_\lambda^{-1}(x))$, where $\sigma(I,\lambda) = \pm 1$ is defined in (5.32)-(5.33).
(iii) *The composition $\Phi'_\nu \circ \rho_\lambda$ is the map*

$$\Phi'_\lambda \colon \prod_{p\in P^*} \mathrm{C}[V_p] \longrightarrow \mathrm{C}[V] \tag{5.34}$$

of (5.26) which can be identified with the operad structure map $\Phi_{\widehat{\lambda}}$ of (5.23).

PROOF. (i) is Lemma 5.6.4 with the orientations given right after it.

For (ii), remember that $\rho_\lambda = \Phi''_\lambda \circ (\rho_0 \times \mathrm{id})$ where Φ''_λ is a homeomorphism by Lemma 5.6.3. When $|P| \geq 2$, Lemma 5.5.2 implies that ρ_0 induces a map of degree ± 1 between the fibers over $\mathrm{C}[P]$, and when $|P| \leq 1$, using that λ is essential, ρ_0 is the identity map. Hence in both cases ρ_λ induces a map of degree ± 1 between the fibers over $\prod_{p \in P^*} \mathrm{C}[A_p]$. We have fixed the orientations so that the fibers of π_λ are oriented according to the linear order of $\oslash_{p \in P^*} I_p$ and the fibers of π'_λ are oriented according to the linear order of I. The number of transpositions needed to reorder $\oslash_{p\in P^*} I_p$ as I is exactly the cardinality of $S(I, \lambda)$. So the sign of the degree of ρ_λ on the fibers is a consequence of the change of orientation rule in Theorem 5.3.2.

For (iii), the equation $\Phi'_\nu \circ \rho_\lambda = \Phi'_\lambda$ follows from the construction of ρ_λ and Φ'_λ. The identification of that map with the operad structure map $\Phi_{\hat{\lambda}}$ is through the canonical homeomorphism

$$\mathrm{C}[V_0] \times \left(\prod_{i \in I_0} \mathrm{C}[\{i\}] \times \prod_{p \in P} \mathrm{C}[V_p] \right) \xrightarrow{\cong} \prod_{p \in P^*} \mathrm{C}[V_p]$$

induced by the projection (see (5.23) and (5.26)). □

Recall from Proposition 5.6.2 the decomposition

$$\mathrm{C}[V, \nu] = \bigcup_{\lambda \in \mathrm{EssCond}(V,\nu)} \mathrm{C}[V, \lambda].$$

We just proved that the fibers $\pi_\lambda'^{-1}(x)$ of each

$$\pi'_\lambda \colon \mathrm{C}[V, \lambda] \to \prod_{p \in P^*} \mathrm{C}[A_p]$$

are of dimension $N \cdot |I|$. Our next proposition can then be interpreted as saying that the pairwise intersections of the terms in this union are of codimension ≥ 1. In other words, the above union is a partition of $\mathrm{C}[V, \nu]$ "up to codimension 1".

PROPOSITION 5.6.6. *If $\lambda \neq \mu$ in $\mathrm{EssCond}(V, \nu)$, then for each $x \in \prod_{p \in P^*} \mathrm{C}[A_p]$,*

$$\dim \left(\pi_\lambda'^{-1}(x) \cap \pi_\mu'^{-1}(x) \right) < N \cdot |I|.$$

PROOF. Let $x \in \prod_{p \in P^*} \mathrm{C}[A_p]$ and pick $v \in I$ such that $\lambda(v) \neq \mu(v)$. For concreteness suppose that $\lambda(v) = k$ where $k = \max(P)$. Set $V_p = V \cap \lambda^{-1}(p)$, for $p \in P$, and $V_0 = \lambda^{-1}(0) \cup P$. Hence $v \in V_k$.

If $(y, x) \in \mathrm{C}[V, \lambda] \cap \mathrm{C}[V, \mu]$ then, as $y \in \mathrm{C}[V]$ is λ-local,

$$\forall a \in A_k, \forall b \in A_p \text{ with } p \neq k : y(a) \simeq y(v) \operatorname{rel} y(b) \tag{5.35}$$

and, as y is also μ-local,

$$\forall a, a' \in A_k : y(a) \simeq y(a') \operatorname{rel} y(v). \tag{5.36}$$

Consider the following diagram in which π'' and π_v are products of canonical projections, $\Phi_2 \colon \mathrm{C}[2] \times \mathrm{C}[A_k] \times \mathrm{C}[\{v\}] \to \mathrm{C}[A_k \cup \{v\}]$ is an operad structure map, and the upper left square is a pullback:

$$
\begin{array}{ccccc}
E & \xrightarrow{\quad\widehat{j}\quad} & \prod_{p\in P^*} \mathrm{C}[V_p] & \xrightarrow{\rho_\lambda} & \mathrm{C}[V,\lambda] \\
\Big\downarrow \widehat{\pi''} & \text{pullback} & \Big\downarrow \pi'' & & \\
\prod_{p<k} \mathrm{C}[A_p] \times (\mathrm{C}[2] \times \mathrm{C}[A_k] \times \mathrm{C}[\{v\}]) & \xhookrightarrow{j:=\mathrm{id}\times\Phi_2} & \prod_{p<k} \mathrm{C}[A_p] \times \mathrm{C}[A_k \cup \{v\}] & & \Big\downarrow \pi'_\lambda \\
& & \Big\downarrow \pi_v & & \\
& & \prod_{p\in P^*} \mathrm{C}[A_p]. & \longleftarrow &
\end{array}
$$

The proximity relations (5.35) and (5.36) imply that

$$\mathrm{C}[V,\lambda] \cap \mathrm{C}[V,\mu] \subset \mathrm{im}(\rho_\lambda \circ \widehat{j}).$$

Therefore

$$
\begin{aligned}
\dim\left(\pi_\lambda'^{-1}(x) \cap \pi_\mu'^{-1}(x)\right) &= \dim(\mathrm{fiber}(\pi'_\lambda) \cap \mathrm{C}[V,\mu]) \\
&\le \dim(\mathrm{fiber}(\pi_v \circ \pi'') \cap \widehat{j}(E)) \\
&= \dim(\mathrm{fiber}(\pi_v \circ j \circ \widehat{\pi''})) \\
&= \dim(\mathrm{fiber}(\pi_v \circ j)) + \dim(\mathrm{fiber}(\widehat{\pi''})) \\
&= \dim(\mathrm{C}[2]) + \dim(\mathrm{fiber}(\pi'')) \\
&= (N-1) + N \cdot (|I| - 1) < N \cdot |I|.
\end{aligned}
$$

□

5.7. Decomposition of the fiberwise boundary along a canonical projection

We turn now to a fiberwise version of the decomposition of the boundary, extending the results of Section 5.4. These results will be needed in Section 9.4 to prove that the Kontsevich configuration spaces integral I is a chain map.

Let $A \subset V$ and consider the canonical projection

$$\pi\colon \mathrm{C}[V] \longrightarrow \mathrm{C}[A]$$

which is a bundle whose fibers are oriented compact manifolds by Theorem 5.3.2 (which we will prove in Section 5.9). Recall from (4.8) and [**18**, Definition 8.1] the fiberwise boundary of an oriented SA bundle. The fiberwise boundary of π is

$$\mathrm{C}^\partial[V] := \bigcup_{x\in \mathrm{C}[A]} \partial(\pi^{-1}(x)) \tag{5.37}$$

which is a closed subspace of $\mathrm{C}[V]$. This space is *not* the same as

$$\partial\, \mathrm{C}[V] \quad \text{or} \quad \bigcup_{x\in \mathrm{C}[A]} \pi^{-1}(x) \cap \partial\, \mathrm{C}[V]$$

(see the example of $[0,1] \times [0,1] \to [0,1]$ right after (4.8)).

We also consider the restriction map

$$\pi^\partial := (\pi|\, \mathrm{C}^\partial[V])\colon \mathrm{C}^\partial[V] \longrightarrow \mathrm{C}[A].$$

Recall from (5.12) in Section 5.4 the set $\mathcal{BF}(V)$ indexing the faces of $\partial\, \mathrm{C}[V]$ and define

$$\mathcal{BF}(V,A) = \{W \in \mathcal{BF}(V) : A \subset W \text{ or } |W \cap A| \le 1\}. \tag{5.38}$$

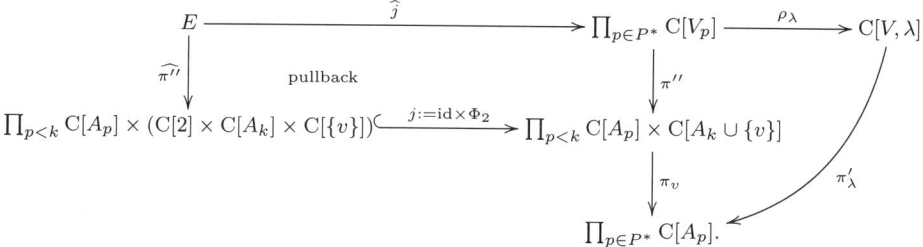

The following is a fiberwise version of Proposition 5.4.1.

PROPOSITION 5.7.1. *There is a decomposition*

$$\mathrm{C}^{\partial}[V] = \bigcup_{W \in \mathcal{BF}(V,A)} \mathrm{im}(\Phi_W)$$

where Φ_W *are the maps defined in* (5.11) *of Section* 5.4.

PROOF. Recall that $\mathrm{C}(A)$ is the interior of the compact manifold $\mathrm{C}[A]$, that is

$$\mathrm{C}(A) = \mathrm{C}[A] \setminus \partial\,\mathrm{C}[A].$$

Then

$$\mathrm{C}^{\partial}[V] \cap \pi^{-1}(\mathrm{C}(A)) \;=\; (\partial\,\mathrm{C}[V]) \cap \pi^{-1}(\mathrm{C}(A)).$$

Since $\mathrm{C}^{\partial}[V]$ is a bundle over $\mathrm{C}[A]$ and $\mathrm{C}[A] = \overline{\mathrm{C}(A)}$, we get that

$$\mathrm{C}^{\partial}[V] = \overline{\mathrm{C}^{\partial}[V] \cap \pi^{-1}(\mathrm{C}(A))} = \overline{(\partial\,\mathrm{C}[V]) \cap \pi^{-1}(\mathrm{C}(A))}$$

where by $\overline{E}$ we mean the topological closure of the subspace E.

For $W \in \mathcal{BF}(V)$, if $A \not\subset W$ and $|W \cap A| \geq 2$, then $\pi(\mathrm{im}\,\Phi_W) \subset \partial\,\mathrm{C}[A]$ because $W \cap A \in \mathcal{BF}(A)$ and $\pi(\mathrm{im}\,\Phi_W)$ is in the image of

$$\Phi^A_{W\cap A}\colon\, \mathrm{C}[A/(W \cap A)] \times \mathrm{C}[W \cap A] \longrightarrow \mathrm{C}[A].$$

Therefore, using Proposition 5.4.1(i),

$$\begin{aligned}
\mathrm{C}^{\partial}[V] &= \overline{\partial\,\mathrm{C}[V] \cap \pi^{-1}(\mathrm{C}(A))} \\
&= \overline{\cup_{W\in\mathcal{BF}(V)}\, \mathrm{im}(\Phi_W) \cap \pi^{-1}(\mathrm{C}(A))} \\
&= \cup_{W\in\mathcal{BF}(V,A)} \overline{\mathrm{im}(\Phi_W) \cap \pi^{-1}(\mathrm{C}(A))} \\
&= \cup_{W\in\mathcal{BF}(V,A)}\, \mathrm{im}(\Phi_W).
\end{aligned}$$

□

5.8. Orientation of $\mathrm{C}[A]$

In this section, we fix an orientation on $\mathrm{C}[A]$. This will be important since we will integrate over this manifold. The orientation will be canonical when N is even and will depend on a linear order on A when N is odd. We will also fix an orientation on the sphere S^{N-1}.

We first review a few classical facts and fix our conventions about orientation:

- A codimension 0 submanifold of an oriented manifold inherits that orientation;
- Conversely, the orientation of a connected manifold is determined by the orientation of any non-empty codimension 0 connected submanifold;
- The product $M_1 \times M_2$ of two oriented manifolds has a canonical orientation. Exchanging the factors preserves or reverses that orientation according to the sign

$$(-1)^{\dim(M_1)\cdot\dim(M_2)};$$

- $\mathbb{R}$, and hence $\mathbb{R}^N = \mathbb{R}\times\cdots\times\mathbb{R}$, is equipped with the standard orientation;

- When M is an oriented smooth manifold and ω is a smooth differential form with compact support of maximal degree on M, one can consider the integral

$$\int_M \omega \in \mathbb{R};$$

- The orientation of a non-empty connected smooth manifold M corresponds to an equivalence class of a smooth differential form α of maximal degree with connected non-empty bounded non-vanishing set, so that $\int_M \alpha > 0$;
- We orient the boundary of a manifold so that the Stokes' formula holds without a sign, that is,

$$\int_{\partial M} \omega = \int_M d\omega$$

for a smooth differential form ω with compact support and of maximal degree on the smooth oriented manifold M.

When $|A| \leq 1$ then $\mathrm{C}[A]$ is a one-point space and we choose the positive orientation on it. Suppose now that $|A| \geq 2$ and suppose given a linear order on A. We then have a natural orientation on the codimension 0 submanifold

$$\mathrm{Inj}(A, \mathbb{R}^N) \subset \prod_{a \in A} \mathbb{R}^N$$

defined in (5.1), where the product is taken in the linear order of A. A transposition in the linear order of A changes this orientation by a sign $(-1)^N$.

Set

$$\mathrm{Inj}_0(A, \mathbb{R}^N) = \left\{x \in \mathrm{Inj}(A, \mathbb{R}^N) : \mathrm{barycenter}(x) = 0\right\}.$$

This is a manifold without boundary. We have a diffeomorphism

$$\begin{array}{ccc} \mathrm{Inj}_0(A, \mathbb{R}^N) \times \mathbb{R}^N & \xrightarrow{\cong} & \mathrm{Inj}(A, \mathbb{R}^N) \\ (x, b) & \longmapsto & x + b \end{array}$$

defined by $(x + b)(a) := x(a) + b$, for $a \in A$. We fix the unique orientation on $\mathrm{Inj}_0(A, \mathbb{R}^N)$ for which the above diffeomorphism preserves the orientation. Consider the codimension 0 submanifold

$$\mathrm{Inj}_0^{\leq 1}(A, \mathbb{R}^N) := \left\{x \in \mathrm{Inj}_0(A, \mathbb{R}^N) : \mathrm{radius}(x) \leq 1\right\} \subset \mathrm{Inj}_0(A, \mathbb{R}^N)$$

with the induced orientation. This is a manifold with boundary and its boundary inherits the orientation.

Identifying $\mathrm{C}(A)$ with $\mathrm{Inj}_0^1(A, \mathbb{R}^N)$ from (5.5), we have

$$\mathrm{C}(A) = \partial\, \mathrm{Inj}_0^{\leq 1}(A, \mathbb{R}^N)$$

and this defines our prefered orientation on $\mathrm{C}(A)$, and hence on $\mathrm{C}[A]$.

We orient the sphere S^{N-1} so that the map

$$\theta_{a,b} \colon \mathrm{C}([\{a, b\}]) \xrightarrow{\cong} S^{N-1}$$

from (5.6) is orientation-preserving when the set $\{a, b\}$ is ordered by $a < b$.

Consider a permutation $\sigma \in \mathrm{Perm}(A)$ of the set A. It induces an obvious automorphism $\mathrm{C}[\sigma]$ of the manifold $\mathrm{C}[A]$. We then have

PROPOSITION 5.8.1. *For a permutation σ of A, the induced homeomorphism*

$$\mathrm{C}[\sigma]\colon \mathrm{C}[A] \longrightarrow \mathrm{C}[A]$$

is orientation-preserving or orientation-reversing according to the sign

$$(\mathrm{sign}(\sigma))^N$$

where $\mathrm{sign}(\sigma) = \pm 1$ *is the signature of the permutation* σ.

5.9. Proof of the local triviality of the canonical projections

The only aim of this long section is to prove Theorem 5.3.2, which asserts that the canonical projection

$$\pi\colon \mathrm{C}[V = A \amalg I] \longrightarrow \mathrm{C}[A]$$

is a semi-algebraic oriented fiber bundle with fibers of prescribed dimension. These fibers should be thought of as a compactification of the configuration space of $|I|$ points in $\mathbb{R}^N$ with $|A|$ points removed. In particular, when I is a singleton, the fiber of π is homeomorphic to a closed ball D^N with $|A|$ disjoint open balls removed.

That the projection $C(V) \to C(A)$ is a bundle is a classical result due to Fadell and Neuwirth [**10**]. The proof for the compactified version is more technical because of the existence of a boundary. Note that, although the spaces $\mathrm{C}[V]$ and $\mathrm{C}[A]$ are smooth manifolds with corners, it is *not true* that $\pi\colon \mathrm{C}[V] \to \mathrm{C}[A]$ is a always a smooth bundle since it is not necessarily a submersion as the following example shows.

EXAMPLE 5.9.1. We now show that $\pi\colon \mathrm{C}[4] \to \mathrm{C}[3]$ is not a smooth bundle. Fix $N = 1$, that is, consider configurations of n points on the real line. In that dimension, $\mathrm{C}[n]$ consists of $n!$ copies of the connected component $\mathrm{C}^{\mathrm{incr}}[n]$ corresponding to configurations $(x_1, \ldots, x_n)$ with $x_1 < \cdots < x_n$. For the remainder of this example we only consider this connected component but drop the superscript incr to simplify notation.

The space $\mathrm{C}[n]$ is exactly the Stasheff associahedron K_{n-2} [**29**]. In particular $\mathrm{C}[3]$ is homeomorphic to the interval $[0, 1]$ and $\mathrm{C}[4]$ is homeomorphic to a pentagon. Label the 4 points of a configuration in $\mathrm{C}[4]$ by $V = \{a, b, c, d\}$ and set $A = \{a, b, c\}$. The five vertices of the pentagon are indexed by all possible way of parenthesizing the product $abcd$ in the most refined way as $(ab)(cd)$, $(a(bc))d$, etc. Each parenthetisation encodes the proximity relations of the points of the configurations $(a, b, c, d) \in \mathrm{C}[4]$, as shown in Figure 5.3.

Each of these five vertices corresponds to a point on the boundary of $\mathrm{C}[4]$. For example, the vertex $a((bc)d)$ corresponds to the limit, as $r \to 0^+$, of the configurations

$$(0, 1 - r - r^2, 1 - r, 1) \in \mathrm{C}(4).$$

Similarly $\mathrm{C}[3]$ is an interval whose endpoints are labeled as $(ab)c$ and $a(bc)$.

A smooth chart of the manifold with corners $\mathrm{C}[4]$ about the point $a((bc)d)$ is given by the unique continuous map

$$f\colon [0, 1) \times [0, 1) \longrightarrow \mathrm{C}[4]$$

whose restriction to $(0, 1) \times (0, 1)$ is defined by the map

$$\begin{aligned}(0, 1) \times (0, 1) &\longrightarrow \mathrm{C}(4) \subset \mathrm{C}[4]\\ (r, s) &\mapsto (0, 1 - r, 1 - r + rs, 1).\end{aligned}$$

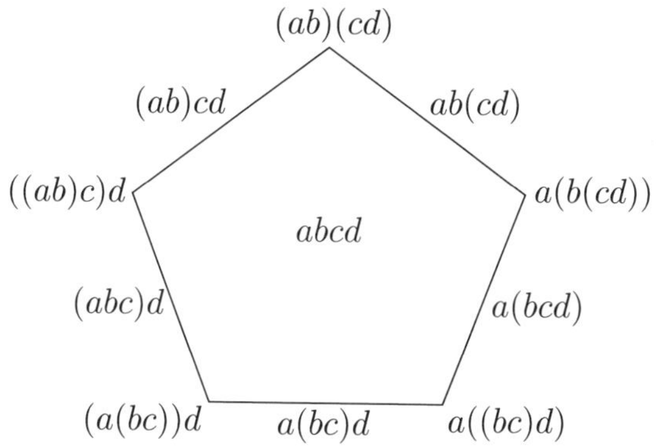

FIGURE 5.3. Stasheff associahedron K_2 depicting the structure of C[4].

Also, there is a chart

$$g\colon (0,1] \longrightarrow \mathrm{C}[3]$$

defined, for $0 < t < 1$, by

$$g(t) = (0, t, 1) \in \mathrm{C}(3) \subset \mathrm{C}[3],$$

and extented continuously to $(0, 1]$.

We then have the following commutative diagram of smooth maps between manifolds with corners

$$\begin{array}{ccc} [0,1) \times [0,1) & \xrightarrow{f} & \mathrm{C}[4] \\ {\scriptstyle p}\downarrow & & \downarrow{\scriptstyle \pi} \\ (0,1] & \xrightarrow{g} & \mathrm{C}[3] \end{array}$$

where

$$p(r,s) = \frac{1-r}{1-r+rs}.$$

The partial derivatives of p are

$$\frac{\partial p}{\partial r}(r,s) = \frac{s}{(1-r+rs)^2}$$

$$\frac{\partial p}{\partial s}(r,s) = \frac{-r(1-r)}{(1-r+rs)^2}.$$

When $r = 0$ and $s = 0$, corresponding to the point $f(0,0) = a((bc)d)$, both these partial derivatives are 0, showing that p is not a submersion at $(0,0)$. Hence π is not a submersion at $a((bc)d)$. Therefore π is not a smooth bundle.

We now come to the proof of Theorem 5.3.2. The composition of two oriented SA bundles is again an oriented SA bundle [**18**, Proposition 8.5], and therefore it is enough to prove that

$$\pi\colon \mathrm{C}[n+1] \longrightarrow \mathrm{C}[n]$$

is an oriented SA bundle. For $n \leq 1$, this is trivial, so we assume that $n \geq 2$. In that case the fiber F of π will be homeomorphic to a disk D^N with n disjoint open disks removed.

We first give a rough idea of the proof in an example. Take $n = 9$ and consider the virtual configuration $x_0 \in \mathrm{C}[9]$ as in Figure 5.4 (see Example 5.2.1 for an explanation of what such a figure represents). We need to build some neighborhood V of x_0 such that the restriction of π over V is equivalent to the projection $V \times F \to V$.

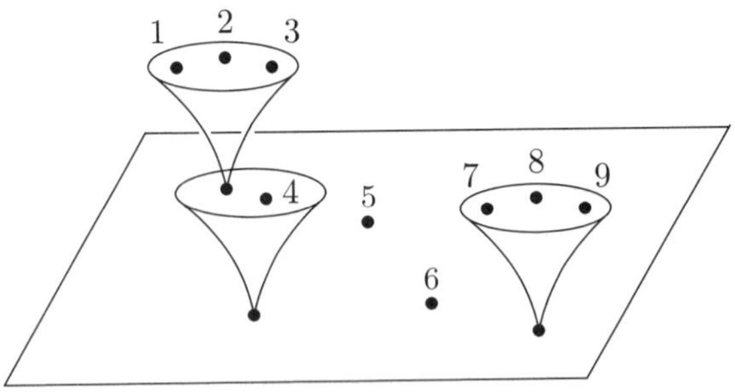

FIGURE 5.4. A virtual configuration $x_0 \in \mathrm{C}[9]$.

For this configuration we have proximity relations such as

$$x_0(1) \simeq x_0(2) \,\mathrm{rel}\, x_0(4),$$
$$x_0(1) \simeq x_0(4) \,\mathrm{rel}\, x_0(5),$$
etc.

All these relations are encoded in the rooted tree T of Figure 5.5.

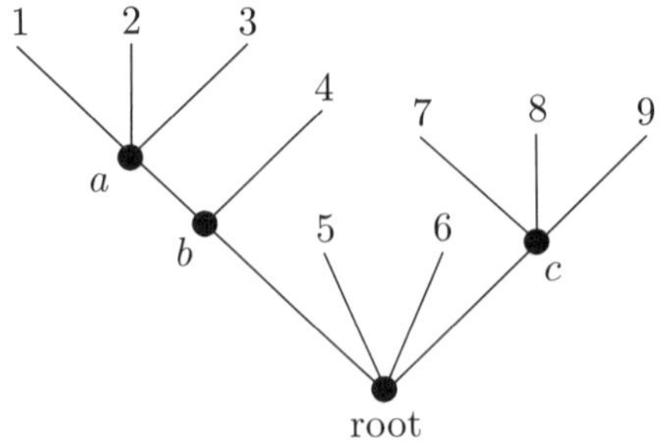

FIGURE 5.5. The tree T associated to $x_0 \in \mathrm{C}[9]$.

To the virtual configuration x_0 we associate a configuration of nested balls in $\mathbb{R}^N$ as in Figure 5.6, with one ball B_v for each vertex $v \in \{1, \ldots, 9, a, b, c, \mathrm{root}\}$ of the tree T, so that $B_v \subset B_w$ iff w is below v in the tree and such that any two balls are either disjoint or one is contained in the other. The centers of the balls labeled by the leaves define a configuration $x_1 = (x_1(1), \ldots, x_1(9)) \in \mathrm{C}(9)$. We also assume that each ball is centered at the barycenter of the centers of the balls immediately contained in that one. For example, B_b is centered at the barycenter of the centers of B_a and B_4. Also, the largest ball B_{root} is centered at the origin.

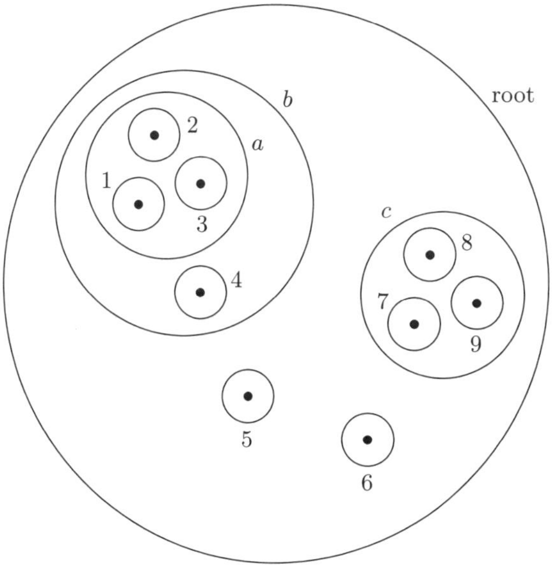

FIGURE 5.6. A configuration of nested balls associated to $x_0 \in \mathrm{C}[9]$.

Consider a self-map

$$\phi_r \colon \mathbb{R}^N \longrightarrow \mathbb{R}^N$$

parametrized by $0 < r \leq 1$ whose effect is to iteratively shrink each ball B_v by a homothety of factor r and extend gradually up to the identity map outside of a small neighborhood of the ball. For $r = 1$, ϕ_r is just the identity, but as $r \to 0$, the image of the configuration x_1 under ϕ_r tends to the virtual configuration x_0.

Now take a point z anywhere inside the outermost closed ball B_{root} but outside of the innermost open balls B_i for $1 \leq i \leq 9$. Let $y_1 = (x_1, z) \in \mathrm{C}(10)$ be the configuration obtained by adjoining the point z to the configuration x_1. Then the image of y_1 under $\lim_{r \to 0} \phi_r$ gives an element in the fiber $\pi^{-1}(x_0) \subset \mathrm{C}[10]$. By choosing the maps ϕ_r with care, we can ensure that the fiber $\pi^{-1}(x_0)$ is covered by such z's, giving a homeomorphism $F \cong \pi^{-1}(x_0)$ where F is a closed ball with 9 small disjoint open balls removed.

We want to prove the local triviality of π, so allow now the centers of the nested balls to move a bit around their initial value while preserving the barycentric relations. Moreover, bound the shrinking of each ball B_v below by some parameter $\tau(v) \in [0, 1]$, for v a vertex other than the root or a leaf $i = 1, \ldots, 9$. Then applying ϕ_r to the configuration of the centers of the balls labeled by the leaves and letting $r \to 0$ describes a neighborhood V of x_0 in $\mathrm{C}[9]$. A parametrized (by V) version of the above construction will then give a trivialization $V \times F \cong \pi^{-1}(V)$ of π over V. This trivialization can be made semi-algebraic and this will prove that π is a semi-algebraic bundle with an oriented compact generic fiber F.

We now proceed with the details of the proof of Theorem 5.3.2. Our goal is to build a neighborhood V of x_0 in $\mathrm{C}[n]$ and a diagram (Diagram (5.59))

$$\begin{array}{ccc} W \times [0, r_1]^{\mathcal{V}_0^*} \times F & \xrightarrow{\widehat{\Phi}} & \mathrm{C}[n+1] \\ \downarrow{\scriptstyle \mathrm{proj}} & & \downarrow{\scriptstyle \pi} \\ W \times [0, r_1]^{\mathcal{V}_0^*} & \underset{\cong}{\xrightarrow{\Phi}} & V \subset \mathrm{C}[n]. \end{array}$$

such that Φ is a semi-algebraic homeomorphism on V and $\widehat{\Phi}$ is a semi-algebraic homeomorphism on $\pi^{-1}(V)$. Here F is the fiber which will be homeomorphic to a unit disk D^N with n open disjoint balls removed. For the domain $W \times [0, r_1]^{\mathcal{V}_0^*}$ of the chart Φ, W is a neighborhood in some products of configuration spaces, and $\mathcal{V}_0^*$ is the set of internal vertices of the tree T associated to x_0.

5.9.1. A stratification of $\mathrm{C}[n]$**.**

We first review a classical stratification of the Fulton-MacPherson configuration spaces indexed by trees (see also [**6**, appendix] and [**27**]).

DEFINITION 5.9.2. A *rooted tree* T *with labels in* $\underline{n} = \{1, \ldots, n\}$ is a tree (that is, an isomorphism class of a simply connected 1-dimensional finite simplicial complex) with one distinguished vertex called the *root* of valence ≥ 2 and such that none of the other vertices is bivalent. The univalent vertices are called the *leaves* and are in bijection with the set $\underline{n}$.

An example is given in Figure 5.5 for $n = 9$.

Denote by $\mathcal{V}$ the set of vertices of the tree T, including the root and the leaves. The leaves are identified with the subset $\underline{n} \subset \mathcal{V}$. Set

$$\mathcal{V}_0 := \mathcal{V} \setminus \{\mathrm{root}\}, \quad \mathcal{V}^* := \mathcal{V} \setminus \underline{n}, \quad \mathcal{V}_0^* := \mathcal{V}^* \cap \mathcal{V}_0.$$

Define a partial order on $\mathcal{V}$ by letting $w \leq v$ when the shortest path in the tree joining v to the root contains w. We write $w < v$ when $w \leq v$ and $w \neq v$. The root is then the minimum of $\mathcal{V}$. Two vertices v_1, v_2 are *not comparable* if neither $v_1 \leq v_2$ nor $v_2 \leq v_1$. For a non-root vertex v we define its *predecessor*

$$\mathrm{pred}(v) := \max\{w \in \mathcal{V} : w < v\}.$$

For a non-leaf vertex w we define its *output set*

$$\mathrm{output}(w) := \{v \in \mathcal{V} : w = \mathrm{pred}(v)\}.$$

The *height function*

$$\mathrm{height}\colon \mathcal{V} \longrightarrow \mathbb{N}$$

is defined inductively by $\mathrm{height}(\mathrm{root}) = 0$ and $\mathrm{height}(v) = \mathrm{height}(\mathrm{pred}(v)) + 1$ when v is not the root.

For example, in the tree of Figure 5.5 we have: $b \leq 1$; $\mathrm{pred}(4) = b$; $\mathrm{output}(\mathrm{root}) = \{b, 5, 6, c\}$; b and 7 are not comparable; and $\mathrm{height}(a) = 2$. For any $w \in \mathcal{V}^*$, $|\,\mathrm{output}(w)| \geq 2$.

For a rooted tree T with leaves labeled by $\underline{n}$ and set of vertices $\mathcal{V}$, consider the product of configuration spaces

$$\mathrm{C}_T := \prod_{w \in \mathcal{V}^*} \mathrm{C}(\mathrm{output}(w)).$$

We now recall how C_T can be identified, via a homeomorphism h_T (see (5.40) below), to a stratum in $C[n]$.Let $\xi = (\xi^w)_{w\in\mathcal{V}^*} \in C_T$. Thus, identifying C(output(w)) with $\mathrm{Inj}_0^1(\mathrm{output}(w), \mathbb{R}^N)$ from (5.5), for $w \in \mathcal{V}^*$ we have

$$\xi^w\colon \mathrm{output}(w) \hookrightarrow \mathbb{R}^N,$$

with

$$\mathrm{barycenter}(\xi^w) = 0 \quad \text{and} \quad \mathrm{radius}(\xi^w) = 1.$$

For $v \in \mathcal{V}_0$ we set

$$\xi(v) := \xi^{\mathrm{pred}(v)}(v).$$

For $r > 0$ and $v \in \mathcal{V}$, define

$$x(\xi, r, v) := \sum_{\substack{w\in\mathcal{V}_0\\ w\leq v}} \xi(w)\cdot r^{\mathrm{height}(w)}. \tag{5.39}$$

The latter formula is equivalent to the inductive definition

$$\begin{cases} x(\xi, r, \mathrm{root}) & = 0 \\ x(\xi, r, v) & = x(\xi, r, \mathrm{pred}(v)) + r^{\mathrm{height}(v)}\xi(v). \end{cases}$$

For $r > 0$ small enough,

$$(x(\xi, r, i)_{1\leq i\leq n})$$

determines a configuration in $C(n)$. When $r \to 0$ this configuration converges to a virtual configuration in $C[n]$ whose proximity relations are described by the tree T. Define

$$h_T\colon C_T \longrightarrow C[n] \tag{5.40}$$

by

$$h_T(\xi) = \lim_{r\to 0+} (x(\xi, r, i))_{1\leq i\leq n}).$$

Then h_T is a homeomorphism onto its image and the family of $\{\mathrm{im}(h_T)\}$, indexed by all rooted trees T with labels in $\underline{n}$, gives a stratification of $C[n]$ [**6**, Appendix] (see also [**27**, Sections 2 and 3]). The maximal stratum is $C(n)$, which is the image of h_{T_0} where T_0 is the tree for which all leaves are of height 1.

A comment about the notation in this section might be in order. Along the rest of the proof, and as it has already appeared above, we will need to consider many configurations in $C(A) = \mathrm{Inj}_0^1(A, \mathbb{R}^N)$ or $C[A]$, for some $A \subset \mathcal{V}$. They will sometimes come with various decorations and arguments, such as

$$\xi^w,\ \xi_0^w,\ x(\xi, r),\ x_0,\ x_1,\ x(\xi, r),\ x(\xi, \tau, r),\ x_1, y, y_1, y_2, \text{ etc.} \tag{5.41}$$

We will also consider the components of these configurations in $\mathbb{R}^N$, such as

$$\xi^w(v),\ \xi(w),\ x(\xi, r, v),\ x(\xi, \tau, r_1, \mathrm{root}),\ \xi_0(i),\ x_1(u), \text{ etc.} \tag{5.42}$$

It might therefore sometimes be confusing whether the notation corresponds to a configuration in $C[A]$ or to one of its components in $\mathbb{R}^N$. As a rule of thumb, the notation will correspond to a point in $\mathbb{R}^N$ when the last argument is a vertex, that is an element of $\mathcal{V}$, like

$$v,\ w,\ i,\ \mathrm{root},\ 1, \ldots, n,\ u,\ p,\ q,\ v_1,\ v_2, \ldots \quad \in \mathcal{V},$$

as in (5.42). Otherwise it will be a configuration, as in (5.41). In particular, in ξ^w the vertex w is a superscript and not an argument, and indeed ξ^w is a configuration but $\xi(v)$ is a point in $\mathbb{R}^N$.

5.9.2. The chart Φ about x_0.

Let $x_0 \in \mathrm{C}[n]$. Our first goal is to build a neighborhood V of x_0 over which π will be trivial and a chart Φ of that neighborhood. We have

$$x_0 = h_T(\xi_0) \tag{5.43}$$

for some tree T and some $\xi_0 \in \mathrm{C}_T$.

For a finite set A of at least two elements and for $\zeta \in \mathrm{Inj}_0^1(A, \mathbb{R}^N) = \mathrm{C}(A)$ (see (5.5)) define

$$\delta(\zeta) := \min\{\|\zeta(a) - \zeta(b)\| : a, b \in A,\ a \neq b\} \in (0, 2].$$

Set

$$r_1 := \frac{1}{4} \min\{\delta(\xi_0^w) : w \in \mathcal{V}^*\}, \tag{5.44}$$

which is a positive number and set

$$W := \{\xi \in \mathrm{C}_T : \forall v \in \mathcal{V}_0, \|\xi(v) - \xi_0(v)\| \leq r_1^{n+1}\}$$

which is a compact neighborhood of ξ_0 in C_T.

Consider now any function

$$\tau\colon \mathcal{V}_0^* \longrightarrow [0, r_1]$$

that we extend to $\mathcal{V}$ by $\tau(\mathrm{root}) = 0$ and $\tau(i) = 0$ for $1 \leq i \leq n$. Define for $\xi \in W$ and $0 \leq r \leq r_1$, by induction on the height of $v \in \mathcal{V}$,

$$\begin{cases} x(\xi, \tau, r, \mathrm{root}) & = 0 \\ x(\xi, \tau, r, v) & = x(\xi, \tau, r, \mathrm{pred}(v)) + \xi(v) \cdot \prod_{\mathrm{root} \leq u < v} \max(r, \tau(u)) \in \mathbb{R}^n \end{cases} \tag{5.45}$$

Note that $x(\xi, \tau, r, w)$ is the barycenter of the points $x(\xi, \tau, r, v)$ for $v \in \mathrm{output}(w)$. Note also that when τ is bounded above by r then $x(\xi, \tau, r, v) = x(\xi, r, v)$ from (5.39).

Finally define

$$\begin{array}{rcl} \Phi\colon W \times [0, r_1]^{\mathcal{V}_0^*} & \longrightarrow & \mathrm{C}[n] \\ (\xi, \tau) & \longmapsto & \lim_{r \to 0+} (x(\xi, \tau, r, i))_{1 \leq i \leq n}. \end{array} \tag{5.46}$$

LEMMA 5.9.3. *Φ is a semi-algebraic homeomorphism onto a compact neighborhood of x_0 in $\mathrm{C}[n]$.*

(Statements similar to this one appear in [**27**], but without the semi-algebraic condition.)

PROOF. Let us first show that Φ is semi-algebraic. The map

$$\begin{array}{rcl} \varphi\colon W \times [0, r_1]^{\mathcal{V}_0^*} \times (0, r_1] & \longrightarrow & \mathrm{C}[n] \times (0, r_1] \\ (\xi, \tau, r) & \longmapsto & ((x(\xi, \tau, r, i)_{1 \leq i \leq n}, r) \end{array}$$

is semi-algebraic, and hence the graph of φ is a semi-algebraic set. This map can be continuously extented to a function

$$\overline{\varphi}\colon W \times [0, r_1]^{\mathcal{V}_0^*} \times [0, r_1] \longrightarrow \mathrm{C}[n] \times [0, r_1]$$

whose restriction to $r = 0$ is the limit function Φ (after projection on the first factor). The graph of $\overline{\varphi}$ is also semi-algebraic as it is the closure of a semi-algebraic set.

Therefore $\overline{\varphi}$, and hence Φ as well, is semi-algebraic. (This argument is analogous to [**5**, Proposition 2.9.1].)

Next we prove the injectivity of Φ. Let y be in the image of Φ, that is

$$y = \lim_{r \to 0+} (x(\xi, \tau, r, i))_{1 \leq i \leq n}.$$

We want to show that we can uniquely determine ξ and τ from y. Define inductively, for $w \in \mathcal{V}$, $y(w)$ as the (virtual) barycenter of the points $y(v)$ for $v \in \text{output}(w)$. Then

$$\xi^w = (y(v) : v \in \text{output}(w))$$

and the function τ can be recovered by comparing the radii of the various sets $\{y(v) : v \in \text{output}(w)\}$, $w \in \mathcal{V}^*$. This proves the injectivity of Φ.

Since the domain of Φ is compact, Φ is a homemorphism onto its image and it is clear that this image is a neighborhood of $x_0 = \Phi(\xi_0, 0)$. $\square$

We denote this compact neighborhood of x_0 in $\mathrm{C}[n]$ by

$$V := \Phi(W \times [0, r_1]^{\mathcal{V}_0^*}). \tag{5.47}$$

5.9.3. Shrinking balls to the limit configurations $\Phi(\xi, \tau)$.
We now build a configuration of nested balls of centers $x_1(\xi, v) \in \mathbb{R}^N$ (for $\xi \in W \subset \mathrm{C}_T$ and v a vertex in $\mathcal{V}$) and of suitable radii $\epsilon(v)$, as well as semi-algebraic self-maps ϕ_r of $\mathbb{R}^N$ which will shrink these balls (ϕ_r will depend on $r > 0$, but also on $\xi \in W$ and $\tau \in [0, r_1]^{\mathcal{V}_0^*}$ not appearing in the notation).

The important features are

(1) Applying the shrinking map ϕ_r to the configuration of centers of innermost balls $(x_1(\xi, i))_{1 \leq i \leq n} \in \mathrm{C}(n)$ gives the configurations $x(\xi, \tau, r, i)_{1 \leq i \leq n}$ which serves, as $r \to 0$, to define the chart Φ in (5.46) (Lemma 5.9.6).
(2) The complement of the innermost balls inside the outermost ball will serve as the fiber of the projection π (this will appear in the next section and will be based on the properties of Lemma 5.9.5 (2).)

We define first the centers $x_1(\xi, v)$ and the radii $\epsilon(v)$ of the balls that we will consider. Suppose given $\xi \in W$ and recall the map x defined in (5.39) and the radius $r_1 > 0$ from (5.44). For $v \in \mathcal{V}$, we set

$$x_1(\xi, v) := x(\xi, r_1, v) \tag{5.48}$$

and

$$\epsilon(v) := 4 \cdot r_1^{\text{height}(v)+1}.$$

The balls $\mathrm{B}[x_1(\xi, v), \epsilon(v)]$ satisfy the following nesting properties:

LEMMA 5.9.4.

(1) If $w < v$ in $\mathcal{V}$ then

$$\mathrm{B}[x_1(\xi, v), \epsilon(v)] \subset \mathrm{B}[x_1(\xi, w), \epsilon(w)/3].$$

(2) If v_1 and v_2 are not comparable in $\mathcal{V}$ then

$$\mathrm{B}[x_1(\xi, v_1), \epsilon(v_1)] \cap \mathrm{B}[x_1(\xi, v_2), \epsilon(v_2)] = \emptyset.$$

PROOF. To simplify notation, we set $r = r_1$ in this proof. Note that $r \leq 1/2$ because $\delta(\xi_0^w) \leq 2$.

For $w < v$,

$$x_1(\xi, v) = x_1(\xi, w) + \sum_{w<u\leq v} \xi(u) \cdot r^{\text{height}(u)}.$$

Therefore

$$\begin{aligned}
\|x_1(\xi, v) - x_1(\xi, w)\| + \epsilon(v) &\leq \sum_{w<u\leq v} \|\xi(u)\| \cdot r^{\text{height}(u)} + 4 \cdot r^{\text{height}(v)+1} \\
&\leq \left(\sup_{w<u\leq v} \|\xi(u)\|\right) \cdot \frac{r^{\text{height}(w)+1}}{1-r} + 4 \cdot r^{\text{height}(v)+1} \\
&\leq r^{\text{height}(w)+1} \left(\frac{1}{1-r} + 4 \cdot r\right) \\
&\leq (4/3) \cdot r^{\text{height}(w)+1} \\
&= \epsilon(w)/3.
\end{aligned}$$

This proves the first part of the lemma.

For the second part, suppose first that v_1 and v_2 have a common predecessor w. Then

$$\begin{aligned}
\|\xi(v_2) - \xi(v_1)\| &\geq \|\xi_0(v_2) - \xi_0(v_1)\| - \|\xi(v_1) - \xi_0(v_1)\| - \|\xi(v_2) - \xi_0(v_2)\| \\
&\geq \delta(\xi_0^w) - 2 \cdot r^{n+1} \\
&\geq 4 \cdot r - 2 \cdot r^{n+1} \\
&> 2 \cdot r.
\end{aligned}$$

Since $\text{height}(v_1) = \text{height}(v_2)$ we get

$$\begin{aligned}
\|x_1(\xi, v_1) - x_1(\xi, v_2)\| &= \|\xi(v_1) - \xi(v_2)\| \cdot r^{\text{height}(v_1)} \\
&> 2 \cdot r \cdot r^{\text{height}(v_1)} \\
&= \epsilon(v_1) + \epsilon(v_2).
\end{aligned}$$

This implies the desired formula when v_1 and v_2 have a common predecessor.

For the general case, since v_1 and v_2 are not comparable, there exists $w_1 \leq v_1$ and $w_2 \leq v_2$ such that w_1 and w_2 have a common predecessor. Therefore $\mathrm{B}[x_1(\xi, w_1), \epsilon(w_1)] \cap \mathrm{B}[x_1(\xi, w_2), \epsilon(w_2)] = \emptyset$. Combining this with the fact that, by the first part of the proposition, $\mathrm{B}[x_1(\xi, v_i), \epsilon(v_i)] \subset \mathrm{B}[x_1(\xi, w_i), \epsilon(w_i)]$, for $i = 1, 2$, we deduce the desired formula. □

We next define a suitable morphism shrinking a given ball.

LEMMA 5.9.5. *There exists a continuous semi-algebraic map*

$$\begin{aligned}
\phi\colon \quad \mathbb{R}^N \times [0,1] \times [0,2] \times \mathbb{R}^N &\longrightarrow \mathbb{R}^N \\
(c, r, \epsilon, x) &\mapsto \phi_r^{c,\epsilon}(x)
\end{aligned}$$

with the following properties:

(1) the map $x \mapsto \phi_r^{c,\epsilon}(x)$
   (a) is radial, centered at c*;*
   (b) is the identity outside of the ball $\mathrm{B}(c, \epsilon)$*;*
   (c) restricts on $\mathrm{B}[c, \epsilon/3]$ *to a homothety of rate* r*;*
   (d) when $r > 0$*, it is a self-homeomorphism of* $\mathbb{R}^N$*;*

(e) *when* $r = 0$, *its restriction to* $\mathbb{R}^N \setminus \mathrm{B}[c, \epsilon/2]$ *is a homeomorphism onto* $\mathbb{R}^N \setminus \{c\}$, *and* $\phi_0^{c,\epsilon}(\mathrm{B}[c, \epsilon/2]) = \{c\}$;

(2) *let* $r > 0$ *and let* $x(1), \dots, x(n)$ *be* $n \geq 2$ *distinct points in* $\mathrm{B}[c, \epsilon/3]$; *then*

(a) $(\phi_r^{c,\epsilon}(x(1)), \dots, \phi_r^{c,\epsilon}(x(n)))$ *determines a configuration in* $\mathrm{C}(n)$ *which does not depend on* r; *hence its limit as* $r \to 0+$ *determines the same configuration;*

(b) *if* z_1, z_2 *are two distinct points in* $\mathrm{B}(c, \epsilon/2)$ *and are different from the* $x(p)$*'s for* $1 \leq p \leq n$, *then*

$$y_i := \lim_{r \to 0+} (\phi_r^{c,\epsilon}(x(1)), \dots, \phi_r^{c,\epsilon}(x(n)), \phi_r^{c,\epsilon}(z_i))$$

determines two different configurations y_1 *and* y_2 *in* $\mathrm{C}(n+1)$ *such that* $y_i(p) \not\simeq y_i(q) \operatorname{rel} y_i(n+1)$ *for* $1 \leq p \neq q \leq n$ *and* $i = 1, 2$;

(c) *if* $z \in \mathbb{R}^N \setminus \mathrm{B}(c, \epsilon/2)$ *then*

$$y := \lim_{r \to 0+} (\phi_r^{c,\epsilon}(x(1)), \dots, \phi_r^{c,\epsilon}(x(n)), \phi_r^{c,\epsilon}(z))$$

determines a configuration in $\mathrm{C}[n+1]$ *such that* $y(p) \simeq y(q) \operatorname{rel} y(n+1)$ *for* $1 \leq p, q \leq n$.

PROOF. The proof consists of explicitly constructing the semi-algebraic function ϕ. Define first a semi-algebraic function

$$\begin{array}{rcl} g\colon [0,1] \times \mathbb{R}_+ & \longrightarrow & [0,1] \\ (r, u) & \longmapsto & g(r, u) \end{array}$$

by

$$g(r,u) = \begin{cases} r, & \text{if } 0 \leq u \leq 1/3; \\ \frac{r}{3-6u}, & \text{if } 1/3 \leq u \leq 1/2 \text{ and } \sqrt{r} \leq 3 - 6u; \\ \sqrt{r}, & \text{if } 1/3 \leq u \leq 1/2 \text{ and } \sqrt{r} \geq 3 - 6u; \\ 2\sqrt{r}(1-u) + 2u - 1, & \text{if } 1/2 \leq u \leq 1; \\ 1, & \text{if } u \geq 1. \end{cases}$$

In other words, the function g is determined by the picture in Figure 5.7 where the curve inside the second rectangle is the parabola $\sqrt{r} = 3 - 6u$.

For $c \in \mathbb{R}^n$, $\epsilon > 0$, and $r \geq 0$ define

$$\begin{array}{rcl} \phi_r^{c,\epsilon}\colon \mathbb{R}^N & \longrightarrow & \mathbb{R}^N \\ x & \longmapsto & \phi_r^{c,\epsilon}(x) = c + (x - c) \cdot g\left(r, \dfrac{\|x - c\|}{\epsilon}\right). \end{array} \tag{5.49}$$

Properties (1a-e) of $x \mapsto \phi_r^{c,\epsilon}(x)$ are then immediate.

(2a) follows from (1c).

(2b-c) are consequences of the following properties of g as $r \to 0$:

- For $u < 1/2$, the map

$$u \mapsto \lim_{r \to 0} \frac{g(r,u)}{r}$$

is the constant 1 over $[0, 1/3]$ and gives a semi-algebraic homeomorphism between $[1/3, 1/2)$ and $[1, +\infty)$.

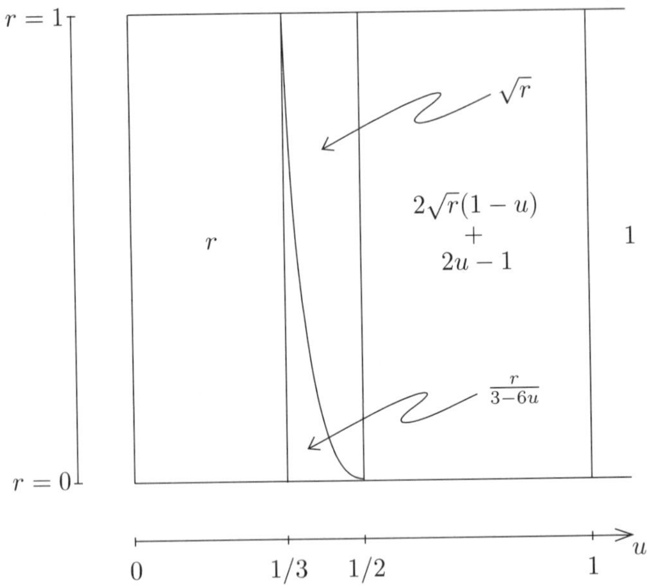

FIGURE 5.7. Definition of the function $(r,u) \longmapsto g(r,u)$.

- For $u \geq 1/2$,
$$\lim_{r\to 0+} \frac{g(r,u)}{r} = +\infty$$
and the map
$$u \mapsto \lim_{r\to 0} g(r,u)$$
is a homeomorphism between $[1/2,1]$ and $[0,1]$.

□

Fix $\xi \in W$ and $\tau \in [0,r_1]^{\mathcal{V}_0^*}$. Recall that we extend τ to $\mathcal{V}$ by 0 on the root and the leaves. For $v \in \mathcal{V}$ and $0 < r \leq r_1$, set
$$\phi_r^v := \phi_{\max(r,\tau(v))/r_1}^{x_1(\xi,v),\epsilon(v)}.$$
Note that ϕ_r^v depends on ξ and τ even if this does not appear in the notation. Then ϕ_r^v is a self-map of $\mathbb{R}^N$ which is the identity outside of $\mathrm{B}[x_1(\xi,v),\epsilon(v)]$, and shrinks the ball $\mathrm{B}[x_1(\xi,v),\epsilon(v)/3]$ by a homothety of rate $\max(r,\tau(v))/r_1$. We will compose all these maps ϕ_r^v for $v \in \mathcal{V}$.

If $v_1, v_2 \in \mathcal{V}$ are two distinct vertices of the same height, then they are non comparable and Lemmas 5.9.5 (1b) and 5.9.4 (2) imply that
$$\phi_r^{v_1} \circ \phi_r^{v_2} = \phi_r^{v_2} \circ \phi_r^{v_1}. \tag{5.50}$$
Let $h_{\max} := \max\{\mathrm{height}(v) : v \in \mathcal{V}\}$. For $h = 1, \ldots, h_{\max}$ we define
$$\phi_r^{[h]} := \mathop{\circ}_{v\in\mathcal{V}_0,\, \mathrm{height}(v)=h} \phi_r^v$$
which is the composition of the maps ϕ_r^v for all vertices v of height h, the order of composition being irrelevant because of (5.50). Finally we set
$$\phi_r := \phi_r^{[1]} \circ \phi_r^{[2]} \circ \cdots \circ \phi_r^{[h_{\max}]}$$

which is a self-map of $\mathbb{R}^N$ which has the effect of iteratively shrinking all the balls of center $x_1(\xi, v)$ starting with the innermost ones first.

Using ϕ_r, we can recover the $x(\xi, \tau, r, i)$, which appears in the chart Φ from (5.46), as follows:

LEMMA 5.9.6. *For $r > 0$ and $1 \leq i \leq n$, we have, in $\mathbb{R}^N$,*

$$\phi_r(x_1(\xi, i)) = x(\xi, \tau, r, i).$$

PROOF. Set, for $w \in \mathcal{V}$,

$$x_r(w) := \begin{cases} \phi_r(x_1(\xi, w)), & \text{if } w \text{ is a leaf;} \\ \text{barycenter}(x_r(v) : v \in \text{output}(w)), & \text{otherwise.} \end{cases} \tag{5.51}$$

Since $x(\xi, \tau, r, v)$ defined in (5.45) satisfies a barycentric relation analoguous to (5.51), the idea of the proof is to compare, by induction on the height of $v \in \mathcal{V}$,

$$x_r(v) \text{ and } x(\xi, \tau, r, v).$$

For the sake of the proof, define for $h \geq 1$ and $w \in \mathcal{V}$

$$\begin{aligned} \phi_r^{\geq[h]} &:= \phi_r^{[h]} \circ \phi_r^{[h+1]} \circ \cdots \circ \phi_r^{[h_{\max}]} \\ \phi_r^{\leq[h]} &:= \phi_r^{[1]} \circ \phi_r^{[2]} \circ \cdots \circ \phi_r^{[h]} \\ x_r^{\geq[h]}(w) &:= \begin{cases} \phi_r^{\geq[h]}(x_1(\xi, w)), & \text{if } w \text{ is a leaf;} \\ \text{barycenter}(x_r^{\geq[h]}(v) : v \in \text{output}(w)), & \text{otherwise.} \end{cases} \end{aligned}$$

In particular, $\phi_r = \phi_r^{\geq[1]}$ and $x_r(v) = x_r^{\geq[1]}(v)$.

We begin by proving three claims on the relations between these self-maps and configurations.

Claim 1: For $w \in \mathcal{V}^*$, $h = \text{height}(w)$, and $v > w$,

$$x_r^{\geq[h]}(v) = \phi_r^w\left(x_r^{\geq[h+1]}(v)\right). \tag{5.52}$$

The claim is proved by induction on $v > w$. If v is a leaf then

$$x_r^{\geq[h]}(v) = \phi_r^{[h]}(x_r^{\geq[h+1]}(v)) = \phi_r^w(x_r^{\geq[h+1]}(v)).$$

Suppose that the claim has been proved when v is replaced by any $u \in \text{output}(v)$ in (5.52). Then

$$\begin{aligned} x_r^{\geq[h]}(v) &= \text{barycenter}(x_r^{\geq[h]}(u) : u \in \text{output}(v)) \\ &\overset{\text{induction}}{=} \text{barycenter}(\phi_r^w(x_r^{\geq[h+1]}(u)) : u \in \text{output}(v)). \end{aligned} \tag{5.53}$$

Set

$$\mathrm{B}_w := \mathrm{B}[x_1(\xi, w), \epsilon(w)/3]. \tag{5.54}$$

By Lemma 5.9.5 (1c), the restriction $\phi_r^w | \mathrm{B}_w$ is a homothety and hence commutes with taking the barycenter. Since $x_r^{\geq[h+1]}(u) \in \mathrm{B}_w$ for $u > w$, we deduce from (5.53) that

$$\begin{aligned} x_r^{\geq[h]}(v) &= \text{barycenter}(\phi_r^w(x_r^{\geq[h+1]}(u)) : u \in \text{output}(v)) \\ &= \phi_r^w(\text{barycenter}(x_r^{\geq[h+1]}(u) : u \in \text{output}(v))) \\ &= \phi_r^w(x_r^{\geq[h+1]}(v)). \end{aligned}$$

This proves Claim 1.

Claim 2: For $v \in \mathcal{V}$,

$$x_r^{\geq[\mathrm{height}(v)]}(v) = x_1(\xi, v).$$

The proof of Claim 2 is by induction. The claim is clear when v is a leaf. Suppose that the claim is true for all $v \in \mathrm{output}(w)$ and set $h = \mathrm{height}(w)$. Then

$$\begin{aligned}
x_r^{\geq[h]}(w) &= \mathrm{barycenter}(x_r^{\geq[h]}(v) : v \in \mathrm{output}(w)) \\
&\overset{\text{Claim 1}}{=} \mathrm{barycenter}(\phi_r^w(x_r^{\geq[h+1]}(v)) : v \in \mathrm{output}(w)) \\
&\overset{\text{induction hyp.}}{=} \mathrm{barycenter}(\phi_r^w(x_1(\xi, v)) : v \in \mathrm{output}(w)) \\
&\overset{\phi_r^w|\mathrm{B}_w \text{ homothety}}{=} \phi_r^w(\mathrm{barycenter}(x_1(\xi, v)) : v \in \mathrm{output}(w)) \\
&= \phi_r^w(x_1(\xi, w)) \\
&= x_1(\xi, w),
\end{aligned}$$

which proves Claim 2.

As a special case we have

$$x_r(\mathrm{root}) = x_1(\xi, \mathrm{root}) = 0. \tag{5.55}$$

Claim 3: For $v \in \mathcal{V}_0$ and $h = \mathrm{height}(v)$,

$$x_r(v) = \phi_r^{\leq[h-1]}(x_1(\xi, v)) = \phi_r^{\leq[h]}(x_1(\xi, v)).$$

Let $w = \mathrm{pred}(v)$. Then the restriction of $\phi_r^{\leq[h-1]}$ to B_w is a composition of homotheties and hence commutes with taking the barycenter. Since $x_r(v)$ is defined by iterated barycenters from a collection of points

$$x_r(i) = \phi_r^{\leq[h-1]}(x_r^{\geq[h]}(i))$$

which belong to the convex B_w (because i are leaves above w), we deduce that

$$\begin{aligned}
x_r(v) &= \phi_r^{\leq[h-1]}(x_r^{\geq[h]}(v)) \\
&\overset{\text{Claim 2}}{=} \phi_r^{\leq[h-1]}(x_1(\xi, v)).
\end{aligned}$$

Finally, since

$$\phi_r^{[h]}(x_1(\xi, v)) = \phi_r^v(x_1(\xi, v)) = x_1(\xi, v)$$

we have

$$\phi_r^{\leq[h-1]}(x_1(\xi, v)) = \phi_r^{\leq[h]}(x_1(\xi, v)).$$

This proves Claim 3.

We are ready for the proof of the lemma. Let $w \in \mathcal{V}^*$. Recall that the restriction of $\phi_r^{\leq[\mathrm{height}(w)]}$ to B_w is a composition of homotheties of total rate

$$R_w := \prod_{\mathrm{root}<u\leq w} \frac{\max(r, \tau(u))}{r_1}.$$

Consider the *normalization* map

$$N : \mathrm{Inj}(A, \mathbb{R}^N) \longrightarrow \mathrm{Inj}_0^1(A, \mathbb{R}^N) = \mathrm{C}[A]$$

that translates the barycenter to the origin and rescales to radius $= 1$. This map is invariant under homotheties of the arguments. Therefore

$$\begin{aligned}
N(x_r(v) : v \in \mathrm{output}(w)) &\overset{\text{Claim 3}}{=} N(\phi_r^{\leq[\mathrm{height}(w)]}(x_1(\xi, v) : v \in \mathrm{output}(w)) \\
&\overset{\text{homotheties}}{=} N(x_1(\xi, v) : v \in \mathrm{output}(w)) \\
&= \xi^w. \qquad (5.56)
\end{aligned}$$

Also

$$\begin{aligned} \operatorname{radius}(x_r(v) : v \in \operatorname{output}(w)) &= \operatorname{radius}(\phi_r^{\leq[\operatorname{height}(w)]}(x_r(v)) : v \in \operatorname{output}(w)) \\ &= R_w \cdot \operatorname{radius}((x_1(\xi, v) : v \in \operatorname{output}(w)) \\ &= R_w \cdot r_1^{\operatorname{height}(w)} \\ &= \prod_{\operatorname{root}<u\leq w} \max(r, \tau(u)). \end{aligned} \tag{5.57}$$

Comparing Equations (5.55), (5.56) and (5.57) with (5.45), we deduce that for all $v \in \mathcal{V}$,

$$x_r(v) = x(\xi, \tau, r, v).$$

The statement of the lemma is the special case when v is a leaf. □

5.9.4. The chart $\widehat{\Phi}$ of $\pi^{-1}(V)$.

We are ready to define the trivialization $\widehat{\Phi}$ of the canonical projection π. Set

$$F := \mathrm{B}[0, n+1] \setminus \cup_{i=1}^{n} \mathrm{B}(i, 1/4).$$

This is a closed ball with n disjoint open balls removed and will serve as the generic fiber of π. For $\xi \in W$, also set

$$F_\xi := \mathrm{B}[x_1(\xi, \operatorname{root}), \epsilon(\operatorname{root})/2] \setminus \cup_{i=1}^{n} \mathrm{B}(x_1(\xi, i), \epsilon(i)/2).$$

It is easy to build semi-algebraic homeomorphisms

$$\Theta_\xi : F \xrightarrow{\cong} F_\xi$$

that depend continuously and semi-algebraically on $\xi \in W$ since W is "small".

Recall the homeomorphism

$$\Phi \colon W \times [0, r_1]^{\mathcal{V}_0^*} \xrightarrow{\cong} V \subset \mathrm{C}[n]$$

from (5.46) and Lemma 5.9.3. Define

$$\widehat{\Phi} \colon W \times [0, r_1]^{\mathcal{V}_0^*} \times F \longrightarrow \mathrm{C}[n+1]$$

by

$$\widehat{\Phi}(\xi, \tau, z_0) := \lim_{r\to 0+} \left(\phi_r(x_1(\xi, 1)), \ldots, \phi_r(x_1(\xi, n)), \phi_r(\Theta_\xi(z_0))\right). \tag{5.58}$$

By (5.46), (5.58), and Lemma 5.9.6, the following diagram commutes:

$$\begin{array}{ccc} W \times [0, r_1]^{\mathcal{V}_0^*} \times F & \xrightarrow{\widehat{\Phi}} & \mathrm{C}[n+1] \\ \downarrow{\scriptstyle \operatorname{proj}} & & \downarrow{\scriptstyle \pi} \\ W \times [0, r_1]^{\mathcal{V}_0^*} & \xrightarrow[\cong]{\Phi} & V \subset \mathrm{C}[n]. \end{array} \tag{5.59}$$

We want to show that $\widehat{\Phi}$ is a homeomorphism onto $\pi^{-1}(V)$, where $V = \operatorname{im}\Phi$ is from (5.47). Fix $(\xi, \tau) \in W \times [0, r_1]^{\mathcal{V}_0^*}$. It is enough to show that $\widehat{\Phi}$ restricts to a homeomorphism on the fibers:

$$\begin{aligned} \hat{\phi} \colon F_\xi &\xrightarrow{\cong} \pi^{-1}(\Phi(\xi, \tau)) \\ z &\longmapsto \widehat{\Phi}(\xi, \tau, \Theta_\xi^{-1}(z)). \end{aligned} \tag{5.60}$$

We first show that $\hat{\phi}$ is injective. Let z_1, z_2 be two distinct elements in F_ξ. Set $y_i = \hat{\phi}(z_i) \in \mathrm{C}[n+1]$ for $i = 1, 2$. We treat different cases.

- Suppose that there exists a vertex $v \in \mathcal{V}$ such that $z_1 \in \mathrm{B}(x_1(\xi, v), \epsilon(v)/2)$ but $z_2 \notin \mathrm{B}(x_1(\xi, v), \epsilon(v)/2)$ (or the other way around). By definition of F_ξ, v is not a leaf. Thus v has at least two distinct outputs and we choose two leaves p and q above each of these outputs. Using Lemma 5.9.5 (2b-c), we get
$$y_1(p) \not\simeq y_1(q) \operatorname{rel} y_1(n+1)$$
but
$$y_2(p) \simeq y_2(q) \operatorname{rel} y_2(n+1).$$
Thus $y_1 \neq y_2$.
- Suppose that the highest vertex $v \in \mathcal{V}$ such that $z_1 \in \mathrm{B}(x_1(\xi, v), \epsilon(v)/2)$ is the same as the highest vertex $w \in \mathcal{V}$ such that $z_2 \in \mathrm{B}(x_1(\xi, w), \epsilon(w)/2)$, that is $v = w$. Choose again two leaves p, q above two distinct outputs of v. Set
$$\phi_r^{\geq v} := \phi_r^{[\mathrm{height}(v)]} \circ \cdots \circ \phi_r^{[\mathrm{height}_{\max}]}.$$
By Lemma 5.9.5 (2b), we have that
$$\lim_{r \to 0} (\phi_r^{\geq v}(z_i), \phi_r^{\geq v}(x_1(\xi, p)), \phi_r^{\geq v}(x_1(\xi, q)))$$
defines two distinct configurations in $\mathrm{C}(3)$. Then applying
$$\lim_{r \to 0} \phi_r^{[0]} \circ \cdots \circ \phi_r^{[\mathrm{height}(v)-1]},$$
which is a composition of homotheties of a ball containing the configurations, still gives two distinct configurations in $\mathrm{C}(3)$. Therefore the images of y_1 and y_2 under some canonical projection $\pi\colon \mathrm{C}[n+1] \to \mathrm{C}[3]$ are distinct. Thus $y_1 \neq y_2$.
- It remains to treat the case when there is no $v \in \mathcal{V}$ such that $z_i \in \mathrm{B}(x_1(\xi, v), \epsilon(v)/2)$ for $i = 1$ or $i = 2$. Then $z_1, z_2 \in \partial\,\mathrm{B}[x_1(\xi, \mathrm{root}), \epsilon(\mathrm{root})/2]$ are in the boundary of the largest ball which is centered at the origin. In that case
$$\theta_{1,n+1}(y_i) = z_i / \|z_i\|$$
where $\theta_{1,n+1}$ from (5.6) gives the direction between the first and the last point of the configuration, and these two directions are distinct. Thus $y_1 \neq y_2$.

This proves that $\hat{\phi}$ is injective. For surjectivity, since F_ξ and $\pi^{-1}(\Phi(\xi, \tau))$ are compact connected manifolds, it is enough to show that $\hat{\phi}$ is surjective on the intersection of the boundary with the fiber. This boundary consists of virtual configurations $y \in \mathrm{C}[n+1]$ such that:

(a) either for some $1 \leq i \leq n$ and for all $j \in \underline{n} \setminus \{i\}$, we have: $y(i) \simeq y(n+1) \operatorname{rel} y(j)$;
(b) or for all $1 \leq i, j \leq n$, we have: $y(i) \simeq y(j) \operatorname{rel} y(n+1)$.

It is clear that $\hat{\phi}$ maps $\partial\,\mathrm{B}[x_1(\xi, i), \epsilon(i)/2]$ surjectively onto the boundaries of type (a) and $\partial\,\mathrm{B}[x_1(\xi, \mathrm{root}), \epsilon(\mathrm{root})/2]$ onto that of type (b).

This proves that $\hat{\phi}$ from (5.60) is a homeomorphism and hence that $\widehat{\Phi}$ in Diagram (5.59) is a homeomorphism onto $\pi^{-1}(V)$. Thus
$$\pi\colon \mathrm{C}[n+1] \longrightarrow \mathrm{C}[n]$$

is an SA bundle. For $n \geq 2$, its fiber F is a compact manifold of dimension N whose interior is homeomorphic to $\mathbb{R}^N$ with n points removed. For $n = 0$ the fiber is a point and for $n = 1$ the fiber is an $(N-1)$-dimensional sphere S^{N-1}.

Thus $\pi : \mathrm{C}[V := A \amalg I] \to \mathrm{C}[A]$ is an oriented SA bundle, as it is the composition of oriented SA bundles $\pi\colon \mathrm{C}[n+1] \to \mathrm{C}[n]$ [**18**, Proposition 8.5]. When $|A| \geq 2$, the interior of the fiber of π can be identified with $\mathrm{Inj}(I, \mathbb{R}^N \setminus A)$ which is of codimension 0 inside $(\mathbb{R}^N)^I$. The latter manifold has a canonical orientation when N is even or when N is odd and I is linearly ordered, and in the second case a transposition in the linear order reverses the orientation.

This finishes the proof of Theorem 5.3.2.

CHAPTER 6

The CDGAs of admissible diagrams

In this chapter we introduce the CDGA of *admissible diagrams* $\mathcal{D}(A)$, where A is a finite set, for example $A = \underline{n} = \{1, \ldots, n\}$. As we will prove later, this differential algebra is a model for both $\Omega_{PA}(\mathrm{C}[A])$ and its cohomology, and it will serve as an intermediate model in the formality proof. In Chapter 7 we will endow $\mathcal{D} := \{\mathcal{D}(n)\}_{n\geq 0}$ with the structure of a cooperad.

The CDGA $\mathcal{D}(A)$ could be defined directly but we will describe it as a quotient of a larger CDGA of *diagrams* $\widehat{\mathcal{D}}(A)$ that we will introduce first. One reason for doing so is that it will be easier to define a cooperad structure on $\widehat{\mathcal{D}} := \{\widehat{\mathcal{D}}(n)\}_{n\geq 0}$ and establish some of its properties, and then induce from this the cooperad structure for $\mathcal{D}$.

In this entire chapter we fix an integer $N \geq 2$ which is the ambient dimension and a unital commutative ring $\mathbb{K}$. The case $N = 1$ is somewhat special, although trivial, and will be treated separately in Chapter 10.

6.1. Diagrams

Roughly speaking, a *diagram* is a finite oriented graph where the vertices come in two flavors, *external* and *internal*, and where the sets of edges and internal vertices are linearly ordered. An example is given in Figure 6.1 and explained in Example 6.1.2 below. The precise definition is as follows.

DEFINITION 6.1.1. A *diagram* is a quintuple $\Gamma = (A_\Gamma, I_\Gamma, E_\Gamma, s_\Gamma, t_\Gamma)$ where

- A_Γ is a finite set;
- I_Γ is a linearly ordered finite set disjoint from A_Γ;
- E_Γ is a linearly ordered finite set; and
- $s_\Gamma, t_\Gamma \colon E_\Gamma \to A_\Gamma \amalg I_\Gamma$ are functions.

We fix the following terminology and notation:

- the elements of A_Γ are the *external vertices*, the elements of I_Γ are the *internal vertices*, and we set $V_\Gamma := A_\Gamma \amalg I_\Gamma$; this is the set of all *vertices*. We extend the order of I_Γ to a partial order on V_Γ by letting $a < i$ when $a \in A_\Gamma$ and $i \in I_\Gamma$;
- the elements of E_Γ are the *edges*;
- $s_\Gamma(e)$ is the *source* and $t_\Gamma(e)$ is the *target* of the edge e; both are the *endpoints* of the edge;
- two distinct vertices are called *adjacent* if they are the endpoints of some edge;
- we say that the edge e is *oriented* from $s_\Gamma(e)$ to $t_\Gamma(e)$;
- we partition the set of edges into the following four families:
  - a *loop* is an edge whose endpoints are identical;
  - a *chord* is an edge between two distinct external vertices;

  - a *dead end* is an edge that is not a loop and such that at least one if its endpoints is internal and has only one adjacent vertex;
  - a *contractible edge* is an edge that is neither a chord, nor a loop, nor a dead end;
- we denote by $E_\Gamma^{\mathrm{contr}}$ the set of contractible edges of Γ;
- the *valence* of a vertex is the number of edges for which the vertex is an endpoint, with loops adding two to the valence;
- an edge e is *simple* if there exists no other edge with the same set of endpoints;
- *double edges* are distinct edges having the same set of endpoints, that is, a pair $\{e_1, e_2\}$ such that $\{s_\Gamma(e_1), t_\Gamma(e_1)\} = \{s_\Gamma(e_2), t_\Gamma(e_2)\}$;
- two vertices v and w are *connected* if there exists a path of edges joining them (ignoring orientations), that is, if there exists a sequence of edges $e_1, \ldots, e_k$ such that $v \in \{s_\Gamma(e_1), t_\Gamma(e_1)\}$, $w \in \{s_\Gamma(e_k), t_\Gamma(e_k)\}$, and $\{s_\Gamma(e_i), t_\Gamma(e_i)\} \cap \{s_\Gamma(e_{i+1}), t_\Gamma(e_{i+1})\} \neq \emptyset$ for $1 \le i < k$;
- given a finite set A, a *diagram on* A is a diagram Γ such that $A_\Gamma = A$;
- a diagram on A is a *unit* if it has no internal vertices or edges. We denote a unit by $\mathbf{1}$. In other words $\mathbf{1} = (A, \emptyset, \emptyset, \emptyset, \emptyset)$;
- two diagrams Γ and Γ' are *isomorphic* if $A_\Gamma = A_{\Gamma'}$ and there exist two order-preserving bijections $\phi_E \colon E_\Gamma \xrightarrow{\cong} E_{\Gamma'}$ and $\phi_I \colon I_\Gamma \xrightarrow{\cong} I_{\Gamma'}$ (that we extend into a bijection $\phi_V := \mathrm{id}_{A_\Gamma} \amalg \phi_I \colon V_\Gamma \xrightarrow{\cong} V_{\Gamma'}$) such that $\phi_V \circ s_\Gamma = s_{\Gamma'} \circ \phi_E$ and $\phi_V \circ t_\Gamma = t_{\Gamma'} \circ \phi_E$.

We will abuse notation by denoting a diagram and its isomorphism class by the same letter Γ.

One should be careful about the definition of a *dead end*. Our definition is not equivalent to saying that a dead end is an edge with a univalent internal vertex. Indeed in Example 6.1.2 and Figure 6.1 below, the edge $(12, 14)_1$ is a dead end (because the vertex 14 is internal and has only 12 as an adjacent vertex), although neither of its endpoints 12 and 14 is of valence 1. However, when a diagram has no loops or double edges, dead ends appear only with univalent internal vertices. The reason for distinguishing dead ends from contractible edges will be given in Remark 6.5.3

EXAMPLE 6.1.2. Consider the diagram in Figure 6.1. By convention, all the external vertices are drawn on a horizontal line which is not a part of the graph. This picture represents a diagram Γ with

- the set of external vertices $A_\Gamma = \{1, \ldots, 5\}$;
- the set of internal vertices $I_\Gamma = \{6, \ldots, 15\}$ with its natural order;
- the set E_Γ consists of eighteen edges, each one oriented from the lower to the higher vertex and ordered as follows (right lexicographic order):

$$\begin{aligned}&(3,4) < (1,6) < (2,6) < (3,7) < (6,7)_1 < (6,7)_2 < (7,8) < \\ &< (8,8) < (8,9) < (4,10) < (5,10) < (11,12) < (11,13) < \\ &< (12,13) < (12,14)_1 < (12,14)_2 < (14,14)_1 < (14,14)_2.\end{aligned}$$

There are three loops, at vertices 14 and 8; three dead ends, $(8,9)$, $(12,14)_1$ and $(12,14)_2$; a chord $(3,4)$; double contractible edges $(6,7)_1$ and $(6,7)_2$; and nine other simple contractible edges. The valence of the vertex 3 is 2, that of 8 is 4, that of 14 is 6, that of 15 is 0, etc.

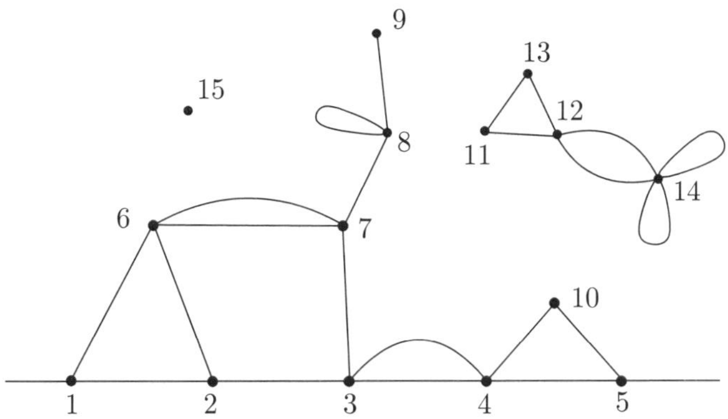

FIGURE 6.1. An example of a diagram (see Example 6.1.2.)

REMARK 6.1.3. Given two diagrams Γ_1 and Γ_2 with the same set of external vertices, we can always find a diagram Γ_2' isomorphic to Γ_2 such that the sets I_{Γ_1} and $I_{\Gamma_2'}$, and E_{Γ_1} and $E_{\Gamma_2'}$ respectively, are disjoint. This will be used in the definition of the product of two (isomorphism classes of) diagrams in Section 6.3. Also, if A and P are disjoint sets and Γ is a diagram on A, we can assume (after maybe replacing diagram Γ by an isomorphic one) that P and I_Γ are disjoint.

6.2. The module $\hat{\mathcal{D}}(A)$ of diagrams

We will define the $\mathbb{K}$-module generated by isomorphism classes of diagrams modulo some signed relations when the linear order of internal vertices or edges is permuted, or the orientation of some edge is reversed. To make this precise, we need the following:

DEFINITION 6.2.1. Let Γ and Γ' be two diagrams with the same set of external vertices.

- Γ and Γ' *differ by an inversion of an edge* if, up to isomorphism, these two diagrams have the same ordered sets of internal vertices and edges, there exists an edge e such that $s_{\Gamma'}(e) = t_\Gamma(e)$ and $t_{\Gamma'}(e) = s_\Gamma(e)$, and s_Γ and $s_{\Gamma'}$ (respectively, t_Γ and $t_{\Gamma'}$) agree on all the other edges.
- Γ and Γ' *differ by a transposition in the linear order of internal vertices*, if, up to isomorphism, they have the same ordered set of edges, the same underlying set of internal vertices I, the same source and target functions, and there exists a transposition $\sigma = (a, b)$ in the group of permutations of the set I, for some pair of distinct internal vertices a and b, such that for all internal vertices $i_1, i_2 \in I$ we have that $i_1 \leq_{I_\Gamma} i_2$ if and only if $\sigma(i_1) \leq_{I_{\Gamma'}} \sigma(i_2)$.
- Γ and Γ' *differ by a transposition in the linear order of the edges*, if, up to isomorphism, they have the same ordered set of internal vertices, the same underlying set of edges E, the same source and target functions, and there exists a transposition $\sigma = (a, b)$ in the group of permutations of the set E, for some pair of distinct edges a and b, such that, for all edges $e_1, e_2 \in E$, $e_1 \leq_{E_\Gamma} e_2$ if and only if $\sigma(e_1) \leq_{E_{\Gamma'}} \sigma(e_2)$.

DEFINITION 6.2.2. Fix an integer $N \geq 1$. The *space of diagrams on a set* A is the free $\mathbb{K}$-module $\widehat{\mathcal{D}}(A)$ generated by the isomorphism classes of diagrams with the set of external vertices A, modulo the equivalence relation $\simeq$ generated by the following:

- $\Gamma \simeq (-1)^N \Gamma'$ if Γ and Γ' differ by an inversion of an edge;
- $\Gamma \simeq (-1)^N \Gamma'$ if Γ and Γ' differ by a transposition in the linear order of internal vertices;
- $\Gamma \simeq (-1)^{N+1} \Gamma'$ if Γ and Γ' differ by a transposition in the linear order of edges.

When we want to emphasize the ambient dimension N, we will denote the space of diagrams by $\widehat{\mathcal{D}}_N(A)$.

By abuse of notation we will denote by the same symbol a diagram and its equivalence class in $\widehat{\mathcal{D}}(A)$.

Because of the relations, when N is odd (respectively even) and $1/2 \in \mathbb{K}$, a diagram with a loop (respectively a double edges) vanishes in the space of diagrams. Other symmetries of a diagram can also make it vanish. Also because of the relations, when N is even the orientation of the edges and the linear order on internal vertices are irrelevant; when N is odd it is the linear order on the edges which is irrelevant. When $1/2 \in \mathbb{K}$, $\widehat{\mathcal{D}}(A)$ is a free $\mathbb{K}-$module generated by a suitable collection of diagrams (if $1/2 \notin \mathbb{K}$ the relation $\simeq$ produces 2-torsion.)

DEFINITION 6.2.3. The *degree* of a diagram Γ is defined to be

$$\deg(\Gamma) = |E_\Gamma| \cdot (N-1) - |I_\Gamma| \cdot N$$

where $|E_\Gamma|$ is the number of edges and $|I_\Gamma|$ is the number of internal vertices.

The motivation for defining $\deg(\Gamma)$ as such comes from the fact that in Chapter 9 we will construct a differential form $\mathrm{I}(\Gamma) \in \Omega_{PA}(\mathrm{C}[A])$ whose degree is exactly that. The integration producing this form is what motivates the signs in Definition 6.2.2.

The degree is compatible with the equivalence relation $\simeq$, and so $\widehat{\mathcal{D}}(A)$ becomes a graded $\mathbb{K}$-module.

6.3. Product of diagrams

Let Γ_1 and Γ_2 be two isomorphism classes of diagrams on the same set A. By Remark 6.1.3, we can assume that the sets $I_{\Gamma_1}$and I_{Γ_2}, and E_{Γ_1} and E_{Γ_2} respectively, are disjoint. Remember the sum of linearly ordered sets $\oslash$ defined in Section 2.2. Define the product diagram $\Gamma = \Gamma_1 \cdot \Gamma_2$ by

- $A_\Gamma := A$;
- $I_\Gamma := I_{\Gamma_1} \oslash I_{\Gamma_2}$;
- $E_\Gamma := E_{\Gamma_1} \oslash E_{\Gamma_2}$;
- $s_\Gamma | E_{\Gamma_i} = s_{\Gamma_i}$ and $t_\Gamma | E_{\Gamma_i} = t_{\Gamma_i}$.

EXAMPLE 6.3.1. An example of a product of two isomorphism classes of diagrams is represented in Figure 6.2. In each picture the edges are oriented from the lower-labeled to the higher-labeled vertex and are ordered by the right lexicographic order as in Example 6.1.2.

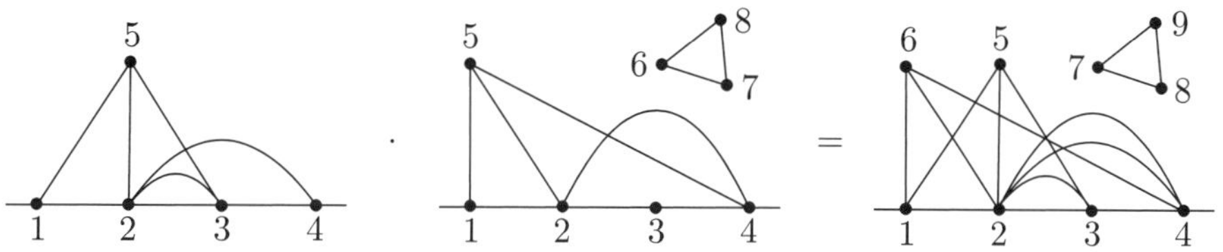

FIGURE 6.2. Example of a product of two diagrams.

PROPOSITION 6.3.2. *The above product extends to a degree* 0 *linear map*

$$\widehat{\mathcal{D}}(A) \otimes \widehat{\mathcal{D}}(A) \longrightarrow \widehat{\mathcal{D}}(A)$$

which endows $\widehat{\mathcal{D}}(A)$ *with the structure of a commutative* $\mathbb{Z}$*-graded algebra.*

PROOF. The multiplication has been defined on generators and we extend it bilinearly. This multiplication is compatible with the equivalence relation $\simeq$ on diagrams. It is also clearly associative and

$$\deg(\Gamma_1 \cdot \Gamma_2) = \deg(\Gamma_1) + \deg(\Gamma_2).$$

The unit diagram $\mathbf{1} = (A, \emptyset, \emptyset, \emptyset, \emptyset)$ is of degree 0 and is indeed a unit for the product.
It remains to check that the multiplication is graded-commutative. Let $\Gamma_i = (A, I_i, E_i, s_i, t_i)$, for $i = 1, 2$, be two diagrams. We distinguish two cases.

- Suppose that N is odd. The diagrams $\Gamma_1 \cdot \Gamma_2$ and $\Gamma_2 \cdot \Gamma_1$ differ by the order of the edges, which is irrelevant in this case, and the order of internal vertices. The number of pairs of transposed vertices is $|I_1| \cdot |I_2|$. Since N is odd, $|I_i| \equiv \deg(\Gamma_i) \mod 2$. Therefore $\Gamma_2 \cdot \Gamma_1 = (-1)^{\deg(\Gamma_1) \cdot \deg(\Gamma_2)} \Gamma_1 \cdot \Gamma_2$.
- Suppose that N is even. The argument is the same as for N odd after exchanging the roles of the linear orders of edges and the internal vertices.

□

6.4. A differential on the space of diagrams

We define now a differential on the $\mathbb{K}$-module $\widehat{\mathcal{D}}(A)$ by "contracting edges" on diagrams. Recall from Definition 6.1.1 the notion of a contractible edge in a diagram. Also remember that in Definition 6.1.1 we extended the linear order on internal vertices into a partial order on the set of all vertices by making external vertices precede internal ones. In particular, if e is a contractible edge of a diagram Γ then the pair $\{s_\Gamma(e), t_\Gamma(e)\}$ is linearly ordered.

DEFINITION 6.4.1. Let Γ be a diagram and let e be a contractible edge of Γ. The diagram obtained from Γ by *contraction of the edge* e is the diagram Γ/e defined as follows:

- $A_{\Gamma/e} := A_\Gamma$
- $I_{\Gamma/e} := I_\Gamma \setminus \{\max(s_\Gamma(e), t_\Gamma(e))\}$
- $E_{\Gamma/e} := E_\Gamma \setminus \{e\}$

- $s_{\Gamma/e} := q \circ s_\Gamma$ and $t_{\Gamma/e} := q \circ t_\Gamma$ where q is defined by:

$$q\colon V_\Gamma \longrightarrow V_{\Gamma/e}$$
$$v \longmapsto \begin{cases} \min(s_\Gamma(e), t_\Gamma(e)), & \text{if } v = \max(s_\Gamma(e), t_\Gamma(e)); \\ v, & \text{otherwise,} \end{cases}$$

where the linear orders on $I_{\Gamma/e}$ and $E_{\Gamma/e}$ are the restrictions of those on I_Γ and E_Γ.

Notice that Γ/e is well-defined because $\max(s_\Gamma(e), t_\Gamma(e))$ is internal since e is not a chord, and $\min(s_\Gamma(e), t_\Gamma(e)) \neq \max(s_\Gamma(e), t_\Gamma(e))$ since e is not a loop.

When e' is an edge distinct from a contractible edge e, we will denote by $\overline{e'}$ the edge of Γ/e corresponding to e' in Γ through the inclusion $E_{\Gamma/e} \hookrightarrow E_\Gamma$.

EXAMPLE 6.4.2. An example of contraction of an edge is given in Figure 6.3 (where we omit precise ordering and orientation of the edges). The edge $\overline{(8,7)}$ is the edge $(8,4)$ in the diagram after contraction of the edge $(4,7)$.

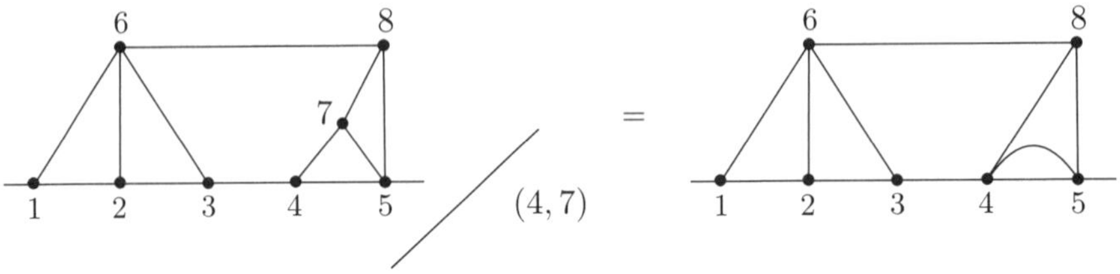

FIGURE 6.3. Contraction of an edge.

Before defining the differential d, we need to introduce some signs. Recall from Section 2.2 the position function pos associated to linearly ordered sets such as I_Γ and E_Γ. Define $\epsilon(\Gamma, e) = \pm 1$ according to the following table (where $s := s_\Gamma$ and $t := t_\Gamma$ are the source and the target of edges)

Value of $\epsilon(\Gamma, e)$	
N odd	N even
$(-1)^{\mathrm{pos}(\max(s(e),t(e)):I_\Gamma)}$ if $s(e) < t(e)$ $-(-1)^{\mathrm{pos}(\max(s(e),t(e)):I_\Gamma)}$ if $s(e) > t(e)$	$(-1)^{\mathrm{pos}(e:E_\Gamma)}$

Let Γ be a diagram on a set of external vertices A. Define its differential $d(\Gamma) \in \widehat{\mathcal{D}}(A)$ by the formula

$$d(\Gamma) := \sum_{e \in E_\Gamma^{\mathrm{contr}}} \epsilon(\Gamma, e) \cdot \Gamma/e \tag{6.1}$$

where the sum runs over all contractible edges e in Γ and the sign $\epsilon(\Gamma, e)$ is from the above table. An example of this is the diagram Γ in Figure 1.2 of the Introduction for which $d(\Gamma)$ is the diagram of Figure 1.1 with $k = 3$, $(i, j, l) = (1, 2, 3)$, and with a suitable orientation and ordering of the edges.

LEMMA 6.4.3. *Formula (6.1) induces a linear map $d\colon \widehat{\mathcal{D}}(A) \to \widehat{\mathcal{D}}(A)$.*

PROOF. The proof that d is compatible with the equivalence relation $\simeq$ of Definition 6.2.2 is straightforward (but tedious). We give the sketch of the argument in one case (the one for which the proof is more complicated) and leave the others to the reader. Proofs of analogous statements can be find in [**8**, Theorem 4.2].

Assume that N is odd and let Γ and Γ' be two diagrams that differ by a transposition in the linear order of internal vertices. Then $\Gamma \simeq -\Gamma'$.

Let a and b be the transposed vertices. Since any transposition of the linear order of internal vertices is obtained as a composition of transpositions of adjacent vertices, we can assume without loss of generality that a and b are consecutive in V_Γ, so

$$\operatorname{pos}(b : I_\Gamma) = \operatorname{pos}(a : I_\Gamma) + 1.$$

Moreover, since it is easy to check that the differential is compatible with inversion of orientations of edges, we can assume that each contractible edge e of Γ is oriented so that $s_\Gamma(e) \leq t_\Gamma(e)$ in V_Γ.

Let e be a contractible edge. We need to show that

$$\epsilon(\Gamma, e) \cdot \Gamma/e \simeq -\epsilon(\Gamma', e) \cdot \Gamma'/e.$$

We distinguish three cases.

(1) Suppose that $\{s_\Gamma(e), t_\Gamma(e)\} = \{a, b\}$. Since $s(e)$ and $t(e)$ are permuted, we have that $\epsilon(\Gamma, e) = -\epsilon(\Gamma', e)$. On the other hand, $\Gamma/e \simeq \Gamma'/e$ since one of the two consecutive vertices a or b disappears.
(2) Suppose that $\{s_\Gamma(e), t_\Gamma(e)\} \cap \{a, b\} = \emptyset$. In that case $\epsilon(\Gamma, e) = \epsilon(\Gamma', e)$ but $\Gamma/e \simeq -\Gamma'/e$.
(3) Suppose that $\{s_\Gamma(e), t_\Gamma(e)\} \cap \{a, b\}$ is a singleton. We have assumed that a and b are consecutive in V_Γ and that $s_\Gamma(e) \leq t_\Gamma(e)$. Moreover $s_\Gamma(e) \neq t_\Gamma(e)$ since e is contractible. Therefore we have four possibilities for the order of $a, b, s_\Gamma(e), t_\Gamma(e)$ in V_Γ:
   (a) $a = s_\Gamma(e) < b < t_\Gamma(e)$. Then $\epsilon(\Gamma, e) = \epsilon(\Gamma', e)$ and $\Gamma/e \simeq -\Gamma'/e$.
   (b) $a < s_\Gamma(e) = b < t_\Gamma(e)$. Then $\epsilon(\Gamma, e) = \epsilon(\Gamma', e)$ and $\Gamma/e \simeq -\Gamma'/e$.
   (c) $s_\Gamma(e) < a < b = t_\Gamma(e)$. Then $\epsilon(\Gamma, e) = -\epsilon(\Gamma', e)$ and $\Gamma/e \simeq \Gamma'/e$.
   (d) $s_\Gamma(e) < a = t_\Gamma(e) < b$. Then $\epsilon(\Gamma, e) = -\epsilon(\Gamma', e)$ and $\Gamma/e \simeq \Gamma'/e$.

In all cases we have $\epsilon(\Gamma, e) \cdot \Gamma/e = -\epsilon(\Gamma', e) \cdot \Gamma'/e$. This proves that $d(\Gamma) = -d(\Gamma')$ as desired. □

LEMMA 6.4.4. *d is homogeneous of degree $+1$.*

PROOF. This is clear from Definition 6.2.3 of degree since, for a contractible edge e of a diagram Γ, the diagram Γ/e has one fewer edge and one fewer internal vertex than Γ. □

LEMMA 6.4.5. *d satisfies the Leibniz rule, that is,*

$$d(\Gamma \cdot \Gamma') = d(\Gamma) \cdot \Gamma' + (-1)^{\deg(\Gamma)} \Gamma \cdot d(\Gamma').$$

PROOF. Recall that $E_{\Gamma \cdot \Gamma'} = E_\Gamma \otimes E_{\Gamma'}$. It is clear than an edge is contractible in Γ or Γ' if and only if it is contractible in $\Gamma \cdot \Gamma'$. Moreover if e is a contractible edge of Γ then $(\Gamma \cdot \Gamma')/e = (\Gamma/e) \cdot \Gamma'$, and if e' is a contractible edge of Γ' then $(\Gamma \cdot \Gamma')/e' = \Gamma \cdot (\Gamma'/e')$. It remains to study the signs ϵ which appear in the differentials, which is straightforward. □

LEMMA 6.4.6. $d^2 = 0$.

PROOF. Let Γ be a diagram and let e_1 and e_2 be distinct edges. If e_1 is contractible, denote by $\overline{e_2}$ the edge in Γ/e_1 corresponding to e_2. It is easy to check that $\overline{e_2}$ is contractible in Γ/e_1 if and only the following two conditions hold:

- e_1 and e_2 are contractible in Γ, and

- e_1 and e_2 do not have the same endpoints, and if e_1 and e_2 have one endpoint in common, then another endpoint of e_1 or e_2 is an internal vertex.

Since these conditions are symmetric, we deduce that $\overline{e_2}$ is contractible in Γ/e_1 if and only if $\overline{e_1}$ is contractible in Γ/e_2, where $\overline{e_1}$ is the edge in Γ/e_2 corresponding to e_1 in Γ. Moreover, in that case $(\Gamma/e_1)/\overline{e_2}$ is isomorphic to $(\Gamma/e_2)/\overline{e_1}$. Therefore

$$d^2(\Gamma) = \sum_{e_1<e_2} \{\epsilon(\Gamma, e_1)\cdot\epsilon(\Gamma/e_1, \overline{e_2}) + \epsilon(\Gamma, e_2)\cdot\epsilon(\Gamma/e_2, \overline{e_1})\} \cdot (\Gamma/e_1)/\overline{e_2}, \tag{6.2}$$

where the sum runs over each pair e_1, e_2 of distinct contractible edges of Γ such that $e_1 < e_2$ and the other condition above making $\overline{e_2}$ contractible in Γ/e_1 holds. It is straightforward to check that the brackets in this sum vanish. □

THEOREM 6.4.7. *$(\widehat{\mathcal{D}}(A), d)$ is a commutative differential $\mathbb{Z}$-graded algebra.*

PROOF. This is a consequence of Proposition 6.3.2 and Lemmas 6.4.3–6.4.6. □

6.5. The CDGA $\mathcal{D}(A)$ of admissible diagrams

DEFINITION 6.5.1. A diagram is *admissible* if it contains no loops, no double edges, no internal vertices of valence ≤ 2, and if each of its internal vertices is connected to some external vertex. Otherwise a diagram is *non-admissible*. We denote by $\mathcal{N}(A)$ the graded submodule of $\widehat{\mathcal{D}}(A)$ generated by the non-admissible diagrams.

An admissible diagram does not have dead ends either, because a dead end implies the existence of an internal vertex of valence 1, or of a loop, or of a double edge. Hence an admissible diagram consists only of simple chords and simple contractible edges.

LEMMA 6.5.2. *The module of non-admissible diagrams $\mathcal{N}(A)$ is a differential ideal of $\widehat{\mathcal{D}}(A)$.*

PROOF. It is easy to check that $\mathcal{N}(A)$ is an ideal of the algebra $\widehat{\mathcal{D}}(A)$.

We show that $\mathcal{N}(A)$ is preserved by the differential d. Let Γ be a non-admissible diagram.

- If Γ contains a loop or a dead end then the same is true for each term of $d(\Gamma)$ since loops and dead ends are not contracted.
- If Γ contains a double edge then each term of $d(\Gamma)$ contains a double edge or a loop (when one of the double edges is contracted).
- If Γ contains a path component with all vertices internal, then the same is true for each term of $d(\Gamma)$.
- If Γ contains an internal vertex i of valence 2 but no double edges or dead ends, then for most of the terms of $d(\Gamma)$, i is still a bivalent internal vertex, except for the two terms obtained by contracting each of the two edges with endpoint i. These two terms cancel each other .
- If Γ has an internal vertex of valence 1 then it has a dead end, and this case is treated in the first bullet above.
- If Γ has an internal vertex of valence 0 then it has a connected component with all vertices internal, and this case is treated in the third bullet above.

This proves that $d(\mathcal{N}(A)) \subset \mathcal{N}(A))$. □

REMARK 6.5.3. The previous lemma would be wrong if in the definition of the differential d we allowed contractions of dead ends. This is why dead ends are not defined as contractible edges in Definition 6.1.1.

DEFINITION 6.5.4. The *$\mathbb{Z}$-graded CDGA of admissible diagrams* is the quotient

$$\mathcal{D}(A) := \widehat{\mathcal{D}}(A)/\mathcal{N}(A).$$

We write $\mathcal{D}_N(A) = \mathcal{D}(A)$ when we want to emphasize the ambient dimension N.

By abuse of notation we will denote by the same symbol a diagram on A, its equivalence class in $\widehat{\mathcal{D}}(A)$, and its larger equivalence class in $\mathcal{D}(A)$. The context should always clear up any ambiguity. As a $\mathbb{K}$-module, $\mathcal{D}(A)$ is generated by admissible diagrams. A (co)chain complex is said to be *connected* if it is concentrated in non-negative degrees and is isomorphic to $\mathbb{K}$ in degree 0.

PROPOSITION 6.5.5. *If $N \geq 3$, then $\mathcal{D}_N(A)$ is a connected CDGA.*

PROOF. Let $\Gamma = (A, I, E, s, t)$ be an admissible diagram different from the unit. We think of an edge of Γ as the union of two half-edges, each with one endpoint which is a vertex of Γ. Since Γ is not the unit and since internal vertices are connected to some external one, there is at least one half-edge whose endpoint is an external vertex. Since each internal vertex is of valence ≥ 3, there are at least $3 \cdot |I|$ other half-edges. Therefore $|E| \geq \frac{1}{2}(1 + 3|I|)$. We deduce that

$$\begin{aligned} \deg(\Gamma) &= |E| \cdot (N-1) - |I| \cdot N \\ &\geq \frac{1}{2}(1+3|I|) \cdot (N-1) - |I| \cdot N \\ &= \frac{N-1}{2} + |I| \cdot \frac{N-3}{2} > 0. \end{aligned}$$

□

REMARK 6.5.6. It is in fact true that $\mathcal{D}_N(A)$ is $(N-3)$-connected for $N \geq 3$. Indeed if $|I| = 0$ then $\deg(\Gamma) = |E| \cdot (N-1) \geq N-1$ and if $|I| \geq 1$ then the inequalities in the above proof show that $\deg(\Gamma) \geq \frac{N-1}{2} + 1 \cdot \frac{N-3}{2} = N-2$. When $N \geq 4$ we can refine this argument by treating separately the cases $|I| = 1$ and $|I| = 2$ and deducing that $\mathcal{D}_N(A)$ is in fact $(N-2)$-connected and of finite type. On the other hand, for $N = 3$ it is not true that it is 1-connected as can be seen from an easy example with $2|E| = 1 + 3|I|$. Also $\mathcal{D}_N(A)$ cannot be $(N-1)$-connected for $|A| \geq 2$ since its homology is the homology of configuration spaces in $\mathbb{R}^N$, as we will see in Theorem 8.1.

However, for $N = 2$, $\mathcal{D}_2(A)$ is not concentrated in non-negative degrees, and is thus not a CDGA which is suitable for modeling a rational homotopy type, even if its cohomology is non-negatively graded (since it is the cohomology of a configuration space).

CHAPTER 7

Cooperad structure on the spaces of (admissible) diagrams

In this chapter we will endow the sequence of CDGAs $\{\mathcal{D}(n)\}_{n\geq 0}$ with the structure of a cooperad. We will do this by first endowing $\{\widehat{\mathcal{D}}(n)\}_{n\geq 0}$ with the structure of a cooperad of graded $\mathbb{K}$-algebras (not differential!). We fix an ambient dimension $N \geq 2$.

The plan is as follows. First we construct in Section 7.1 the cooperad structure maps $\widehat{\Psi}$ and Ψ on $\widehat{\mathcal{D}}$ and $\mathcal{D}$ using the notion of *condensation* from Definition 5.6.1. Then we prove in Section 7.2 that these are morphisms of algebras, and in Section 7.3 we show that Ψ is a chain map (this is not the case for $\widehat{\Psi}$.) Finally we prove in Cection 7.4 that this defines the structure of a cooperad of CDGAs on $\mathcal{D}$; this is our main result, Theorem 7.4.3.

For the several following sections, fix a weak ordered partition $\nu\colon A \to P$ and set $P^* = \{0\} \otimes P$, $A_p = \nu^{-1}(p)$, and $A_0 = P$, as in the setting 2.4.1.

7.1. Construction of the cooperad structure maps $\hat{\Psi}_\nu$ and Ψ_ν

In this section we build maps

$$\widehat{\Psi}_\nu\colon \widehat{\mathcal{D}}(A) \longrightarrow \widehat{\mathcal{D}}(P) \otimes \bigotimes_{p\in P} \widehat{\mathcal{D}}(A_p) \qquad \text{and}$$
$$\Psi_\nu\colon \mathcal{D}(A) \longrightarrow \mathcal{D}(P) \otimes \bigotimes_{p\in P} \mathcal{D}(A_p)$$

which will serve as cooperad structure maps. Of course, the tensor product over $p \in P$ is taken in the order fixed on P. Since $A_0 = P$ we have

$$\widehat{\mathcal{D}}(P) \otimes \bigotimes_{p\in P} \widehat{\mathcal{D}}(A_p) = \bigotimes_{p\in P^*} \widehat{\mathcal{D}}(A_p).$$

Let us first describe roughly the idea of $\widehat{\Psi}_\nu$. Let Γ be a diagram on A with the set of vertices V_Γ. Recall from Definition 5.6.1 that a condensation of V_Γ relative to ν is a map

$$\lambda\colon V_\Gamma \to P^*$$

such that $\lambda|A = \nu$. For each condensation $\lambda \in \mathrm{Cond}(V_\Gamma, \nu)$, we first construct diagrams $\Gamma(\lambda, 0) \in \widehat{\mathcal{D}}(P)$ and $\Gamma(\lambda, p) \in \widehat{\mathcal{D}}(A_p)$ as follows. For $p \in P$, the diagram $\Gamma(\lambda, p)$ on A_p is the full subgraph of Γ whose vertices are the p-locals (that is, those in $\lambda^{-1}(p)$). The diagram $\Gamma(\lambda, 0)$ on P is obtained from Γ by shrinking each subgraph $\Gamma(\lambda, p)$ into a single external vertex p, for $p \in P$. Then $\widehat{\Psi}_\nu(\Gamma) \in \widehat{\mathcal{D}}(P)\otimes\otimes_{p\in P}\widehat{\mathcal{D}}(A_p)$ is defined to be

$$\widehat{\Psi}_\nu(\Gamma) = \sum_{\lambda\in\mathrm{Cond}(V_\Gamma,\nu)} \Gamma(\lambda) \qquad \text{where} \qquad \Gamma(\lambda) = \pm\Gamma(\lambda, 0) \otimes \bigotimes_{p\in P} \Gamma(\lambda, p).$$

The precise formulas are in equations (7.5) and (7.6).

Here is an example illustrating this.

EXAMPLE 7.1.1. Suppose $A = \{1, 2, 3, 4, 5\}$ and $I = \{6, 7\}$, so $V = \{1, 2, 3, 4, 5, 6, 7\}$. Let Γ be as in Figure 7.1.

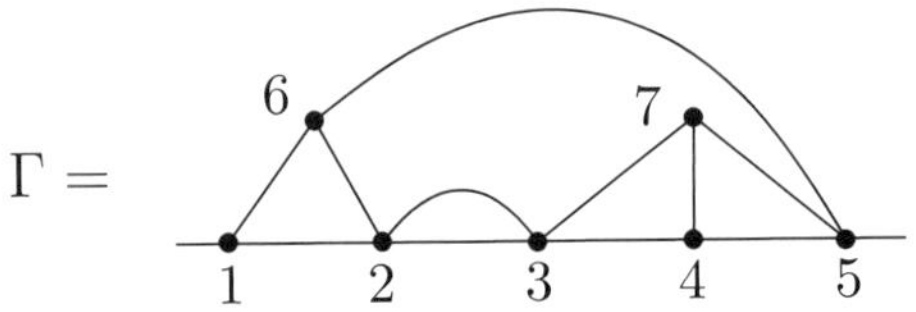

FIGURE 7.1

Let $P^* = \{0, \alpha, \beta\}$ and let $\lambda\colon V \to P^*$ be defined by

$$\lambda(v) = \begin{cases} \alpha, & \text{for } v = 1, 2, 3; \\ \beta, & \text{for } v = 4, 5, 7; \\ 0, & \text{for } v = 6. \end{cases}$$

Then $\Gamma(\lambda, 0)$, $\Gamma(\lambda, \alpha)$, and $\Gamma(\lambda, \beta)$ are given in Figure 7.2.

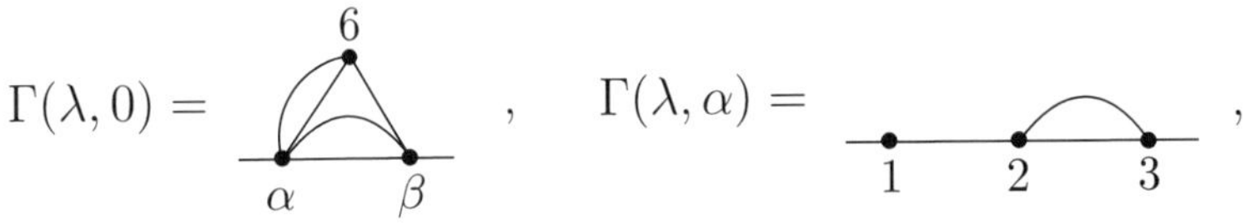

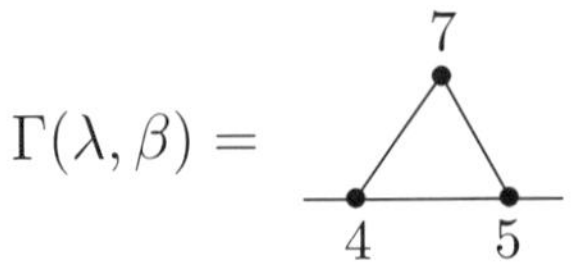

FIGURE 7.2

Here is another heuristic description of $\Gamma(\lambda)$. Picture Γ as in Figure 7.3, so that all the p-local vertices and their connecting edges are drawn infinitesimally close to each other, and all the global vertices and the various clusters of p-local vertices are drawn far from each other. Then $\Gamma(\lambda, 0)$ is the diagram Γ seen from far away, and each $\Gamma(\lambda, p)$ for $p \in P$ is that diagram seen through a microscope centered at the pth cluster (forgeting the edges outside of the cluster). This interpretation will correspond, through the Kontsevich configuration space integral, to what happens to the configurations of points in the Fulton-MacPherson operad. See the discussion after (9.27) as to why the Kontsevich configuration space integral commutes with the cooperadic structures.

DEFINITION 7.1.2. Let $\nu\colon A \to P$ be a weak ordered partition, let $P^* = \{0\} \oslash P$, let Γ be a diagram on A and assume that $I_\Gamma \cap P = \emptyset$.

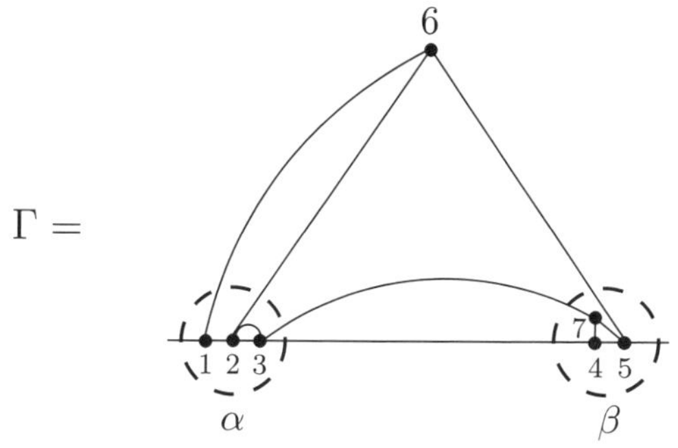

FIGURE 7.3

- A *condensation* λ on Γ is a condensation of V_Γ relative to ν as in Definition 5.6.1, that is, it is a map $\lambda\colon V_\Gamma \to P^*$ such that $\lambda|A = \nu$. We consider the set $\mathrm{Cond}(\Gamma,\nu) := \mathrm{Cond}(V_\Gamma,\nu)$ of all condensations on Γ relative to ν, and write $\mathrm{Cond}(\Gamma)$ when ν is understood.
- The *extension to the edges* of the condensation λ on Γ is the map
$$\lambda_E\colon E_\Gamma \longrightarrow P^*$$
defined by
$$\lambda_E(e) = \begin{cases} \lambda(s_\Gamma(e)), & \text{if } \lambda(s_\Gamma(e)) = \lambda(t_\Gamma(e)), \\ 0, & \text{otherwise.} \end{cases}$$
- Given a condensation λ of Γ, a vertex v (respectively an edge e) is *p-local*, for $p \in P$, if $\lambda(v) = p$ (respectively $\lambda_E(e) = p$). It is *global* if $\lambda(v) = 0$ (respectively $\lambda_E(e) = 0$).

The terminology *condensation* is motivated in the case of diagrams (as it was in Definition 5.6.1 for configurations) by the idea explained right before Definition 7.1.2 that the diagram should be pictured as in Figure 7.3 with its vertices condensed into clusters depending on the values of λ.

Clearly the set of condensations on Γ is in bijective correspondence with the set of maps from I_Γ to P^*, since the value of a condensation λ on an external vertex a is determined by $\lambda(a) = p$ for $a \in A_p$, that is, $\lambda(a) = \nu(a)$. An edge is p-local if and only if both of its endpoints are. Otherwise it is global. Also, a global vertex is always internal but a global edge can be a chord.

Let Γ be a diagram on A and let $\lambda \in \mathrm{Cond}(\Gamma)$. Without loss of generality we can assume that $I_\Gamma \cap P = \emptyset$ (see Remark 6.1.3). For $p \in P^*$ we define a diagram

$$\Gamma(\lambda, p) := (A_p, I_p, E_p, s_p, t_p) \tag{7.1}$$

with

- $I_p = I_\Gamma \cap \lambda^{-1}(p)$;
- $E_p = \lambda_E^{-1}(p)$;
-
  - For $p \in P$, s_p and t_p are the restrictions of s_Γ and t_Γ to E_p;
  - For $p = 0$, $s_0 = \widehat{\lambda} \circ s_\Gamma$ and $t_0 = \widehat{\lambda} \circ t_\Gamma$ where
  $$\widehat{\lambda}\colon V_\Gamma \longrightarrow P \cup I_0$$
  is defined by $\widehat{\lambda}(v) = v$ if $\lambda(v) = 0$, and $\widehat{\lambda}(v) = \lambda(v)$ otherwise.

The set of edges (respectively of internal vertices) of Γ is the disjoint union for $p \in P^*$ of the set of edges (respectively of internal vertices) of the diagrams $\Gamma(\lambda, p)$. Even if Γ is admissible, $\Gamma(\lambda, p)$ may not be. Note also that $\widehat{\lambda}$ above is the same as in (5.22).

The equivalence class of $\Gamma(\lambda, p)$ in $\widehat{\mathcal{D}}(A_p)$ with respect to the relation $\simeq$ of Definition 6.2.2, or even $\otimes_{p\in P^*}\Gamma(\lambda, p)$, is *not* an invariant of the equivalence class of Γ in $\widehat{\mathcal{D}}(A)$. To correct this we introduce some signs. Define for $I = I_\Gamma$ and $E = E_\Gamma$,

$$S(I, \lambda) := \{(v, w) \in I \times I : v < w \text{ and } \lambda(v) > \lambda(w)\}$$

(which was already introduced in (5.33)), and

$$S(E, \lambda) := \{(e, f) \in E \times E : e < f \text{ and } \lambda_E(e) > \lambda_E(f)\}.$$

Define the signs

$$\sigma(I, \lambda) := (-1)^{N\cdot|S(I,\lambda)|}, \tag{7.2}$$

$$\sigma(E, \lambda) := (-1)^{(N-1)\cdot|S(E,\lambda)|}, \text{ and} \tag{7.3}$$

$$\sigma(\Gamma, \lambda) := \sigma(I, \lambda) \cdot \sigma(E, \lambda) \tag{7.4}$$

(The sign $\sigma(I, \lambda)$ was already defined in (5.32).) The proof of the following is straightforward.

LEMMA 7.1.3. *For a diagram Γ and a condensation λ on Γ, the element*

$$\sigma(\Gamma, \lambda) \cdot \bigotimes_{p\in P*} \Gamma(\lambda, p) \quad \in \quad \bigotimes_{p\in P*} \widehat{\mathcal{D}}(A_p)$$

depends only on the equivalence class of Γ in $\widehat{\mathcal{D}}(A)$.

For a diagram Γ on A and a condensation λ of Γ we set

$$\Gamma(\lambda) := \sigma(\Gamma, \lambda) \cdot \bigotimes_{p\in P*} \Gamma(\lambda, p) \quad \in \quad \bigotimes_{p\in P*} \widehat{\mathcal{D}}(A_p), \tag{7.5}$$

where $\sigma(\Gamma, \lambda) = \pm 1$ is from (7.4) and $\Gamma(\lambda, p)$ is from (7.1). By Lemma 7.1.3 we get a linear map

$$\widehat{\Psi}_\nu : \widehat{\mathcal{D}}(A) \longrightarrow \bigotimes_{p\in P*} \widehat{\mathcal{D}}(A_p)$$

defined on generators by

$$\widehat{\Psi}_\nu(\Gamma) := \sum_{\lambda\in \mathrm{Cond}(\Gamma,\nu)} \Gamma(\lambda). \tag{7.6}$$

Recall $\mathcal{N}(A_p) \subset \widehat{\mathcal{D}}(A_p)$, the differential ideal of non-admissible diagrams (Definition 6.5.1 and Lemma 6.5.2). Set

$$\mathcal{N}(\nu) := \sum_{p\in P^*} \bigotimes_{\substack{q\in P^* \\ q<p}} \widehat{\mathcal{D}}(A_q) \otimes \mathcal{N}(A_p) \otimes \bigotimes_{\substack{q\in P^* \\ q>p}} \widehat{\mathcal{D}}(A_q), \tag{7.7}$$

which is a differential ideal in $\otimes_{p\in P^*}\widehat{\mathcal{D}}(A_p)$. Since $\mathcal{D}(A_p) = \widehat{\mathcal{D}}(A_p)/\mathcal{N}(A_p)$, we have an isomorphism of CDGAs ($\mathbb{Z}$-graded if $N = 2$)

$$\left(\bigotimes_{p\in P*} \widehat{\mathcal{D}}(A_p)\right) / \mathcal{N}(\nu) \cong \bigotimes_{p\in P*} \mathcal{D}(A_p) \tag{7.8}$$

LEMMA 7.1.4. $\widehat{\Psi}_\nu(\mathcal{N}(A)) \subset \mathcal{N}(\nu)$.

PROOF. Let Γ be a non-admissible diagram on A and let $\lambda \in \mathrm{Cond}(\Gamma)$.

- If Γ has a loop at a vertex v, then $\Gamma(\lambda, \lambda(v))$ also has a loop.
- If Γ has double edges e_1 and e_2, then so does $\Gamma(\lambda, \lambda_E(e_1))$.
- If Γ has an internal vertex v of valence ≤ 2, then the same is true for $\Gamma(\lambda, \lambda(v))$ because the valence of v can only decrease.
- If, for some $p \in P$, Γ has an internal p-local vertex that is not connected to any external vertex, then the same is true for $\Gamma(\lambda, p)$. If Γ has a connected component consisting only of internal global vertices, then the same is true for $\Gamma(\lambda, 0)$.

In all cases we see that if Γ is not admissible, then the same is true for $\Gamma(\lambda, p)$ for some $p \in P^*$. Therefore $\Gamma(\lambda) \in \mathcal{N}(\nu)$ and $\widehat{\Psi}_\nu(\mathcal{N}(A)) \subset \mathcal{N}(\nu)$. □

PROPOSITION 7.1.5. *$\widehat{\Psi}_\nu$ defined in (7.6) induces a linear map*

$$\Psi_\nu \colon \mathcal{D}(A) \longrightarrow \mathcal{D}(P) \otimes \bigotimes_{p\in P} \mathcal{D}(A_p).$$

PROOF. This is an immediate consequence of the isomorphism (7.8) and Lemma 7.1.4. □

Thus, for an admissible diagram Γ, $\Psi_\nu(\Gamma)$ is obtained as the sum (7.6) in which non-admissible terms are set to zero. Actually, there are many condensations λ for which $\Gamma(\lambda)$ is not admissible and therefore does not contribute to $\Psi_\nu(\Gamma)$. In particular, only *admissible condensations* (to be defined in Definition 7.3.3) can contribute to the sum, and hence we can use the sum (7.11) below, which has many fewer terms, to define $\Psi_\nu(\Gamma)$.

7.2. $\hat{\Psi}_\nu$ and Ψ_ν are morphisms of algebras

The aim of this section is to prove

PROPOSITION 7.2.1. *$\widehat{\Psi}_\nu$ and Ψ_ν are morphisms of algebras.*

PROOF. We first prove the statement for $\widehat{\Psi}_\nu$. Let Γ_1 and Γ_2 be two diagrams on A and suppose that I_{Γ_1} and I_{Γ_2}, E_{Γ_1} and E_{Γ_2} respectively, are disjoint.

Define the function

$$\operatorname{Cond}(\Gamma_1) \times \operatorname{Cond}(\Gamma_2) \longrightarrow \operatorname{Cond}(\Gamma_1 \cdot \Gamma_2), \qquad (\lambda_1, \lambda_2) \longmapsto \lambda_1 \cdot \lambda_2$$

by $(\lambda_1 \cdot \lambda_2)(v) = \lambda_i(v)$ when $v \in V_{\Gamma_i}$ for $i = 1, 2$. This map is well-defined because if $v \in V_{\Gamma_1} \cap V_{\Gamma_2}$ then v is external and $\lambda_1(v) = \lambda_2(v) = \nu(v)$. Moreover, it is a bijection whose inverse is given by $\lambda \mapsto (\lambda|V_{\Gamma_1}, \lambda|V_{\Gamma_2})$.

Since

$$\Gamma_1(\lambda_1, p) \cdot \Gamma_2(\lambda_2, p) = (\Gamma_1 \cdot \Gamma_2)(\lambda_1 \cdot \lambda_2, p)$$

it is easy to see that

$$\bigotimes_{p\in P*} (\Gamma_1 \cdot \Gamma_2)(\lambda_1 \cdot \lambda_2, p) = \eta(\Gamma_1, \lambda_1, \Gamma_2, \lambda_2) \cdot \left(\bigotimes_{p\in P*} \Gamma_1(\lambda_1, p) \right) \cdot \left(\bigotimes_{q\in P*} \Gamma_2(\lambda_2, q) \right).$$

where

$$\eta(\Gamma_1, \lambda_1, \Gamma_2, \lambda_2) := (-1)^s \quad \text{with} \quad s = \sum_{\substack{p,q\in P^* \\ q<p}} \deg(\Gamma_1(\lambda_1, p)) \cdot \deg(\Gamma_2(\lambda_2, q)).$$

We have

$$
\begin{aligned}
\widehat{\Psi}_\nu(\Gamma_1\cdot\Gamma_2) &= \sum_{\lambda\in\mathrm{Cond}(\Gamma_1\cdot\Gamma_2)} \sigma(\Gamma_1\cdot\Gamma_2,\lambda)\cdot \bigotimes_{p\in P*}(\Gamma_1\cdot\Gamma_2)(\lambda,p)\\
&= \sum_{\lambda_1\in\mathrm{Cond}(\Gamma_1)}\sum_{\lambda_2\in\mathrm{Cond}(\Gamma_2)} \sigma(\Gamma_1\cdot\Gamma_2,\lambda_1\cdot\lambda_2)\cdot \bigotimes_{p\in P*}(\Gamma_1\cdot\Gamma_2)(\lambda_1\cdot\lambda_2,p)\\
&= \sum_{\lambda_1\in\mathrm{Cond}(\Gamma_1)}\sum_{\lambda_2\in\mathrm{Cond}(\Gamma_2)} \Big\{\sigma(\Gamma_1\cdot\Gamma_2,\lambda_1\cdot\lambda_2)\cdot\eta(\Gamma_1,\lambda_1,\Gamma_2,\lambda_2)\\
&\qquad\cdot\Big(\bigotimes_{p\in P*}\Gamma_1(\lambda_1,p)\Big)\cdot\Big(\bigotimes_{q\in P*}\Gamma_2(\lambda_2,q)\Big)\Big\}
\end{aligned}\tag{7.9}
$$

On the other hand

$$
\begin{aligned}
\widehat{\Psi}_\nu(\Gamma_1)\cdot\widehat{\Psi}_\nu(\Gamma_2) &= \sum_{\lambda_1\in\mathrm{Cond}(\Gamma_1)}\sum_{\lambda_2\in\mathrm{Cond}(\Gamma_2)} \Gamma_1(\lambda_1)\cdot\Gamma_2(\lambda_2)\\
&= \sum_{\lambda_1\in\mathrm{Cond}(\Gamma_1)}\sum_{\lambda_2\in\mathrm{Cond}(\Gamma_2)} \Big\{\sigma(\Gamma_1,\lambda_1)\cdot\sigma(\Gamma_2,\lambda_2)\\
&\qquad\cdot\Big(\bigotimes_{p\in P*}\Gamma_1(\lambda_1,p)\Big)\cdot\Big(\bigotimes_{q\in P*}\Gamma_2(\lambda_2,q)\Big)\Big\}
\end{aligned}\tag{7.10}
$$

It remains to check that the signs of (7.9) and (7.10) agree, which is straightforward..

For Ψ_ν, the statement is a consequence of the definition of Ψ_ν in Proposition 7.1.5 and of the fact that (7.8) is an isomorphism of algebras. □

7.3. Ψ_ν is a chain map

This section is devoted to the proof of the following

PROPOSITION 7.3.1. *Ψ_ν commutes with the differentials.*

The analog for $\widehat{\Psi}_\nu$ is *not* true, as illustrated in the following example.

EXAMPLE 7.3.2. Here we will show that $\widehat{\Psi}_\nu$ is not a chain map. Consider the diagram Γ given in Figure 7.4, with the set of external vertices $A=\{a\}$. The internal vertices $\{1,2,3,4\}$ have their natural order and each edge is oriented from the lower to the higher vertex. Suppose also that N is odd, and hence the order of edges is irrelevant. Set $P=\{\alpha\}$ and consider the unique ordered partition

$$\nu\colon \{a\}\longrightarrow\{\alpha\}.$$

The boundary of Γ is given by the diagram pictured on the top right of Figure 7.4 (obtained as an alternating sum of three such diagrams).

Then $\Psi_\nu(d\Gamma)$ is a sum of eight terms corresponding to the eight condensations

$$\lambda\colon\{1,2,3\}\longrightarrow P^*=\{0,\alpha\}.$$

Using that dead ends are not contractible and diagrams with loops vanish when N is odd (because of the relations $\simeq$ of Definition 6.2.2), one computes that

$$d(\widehat{\Psi}_\nu(d\Gamma))\in\widehat{\mathcal{D}}(\{\alpha\})\otimes\widehat{\mathcal{D}}(\{a\})$$

consists of a single term, corresponding to the condensation $\lambda(1)=\alpha$, $\lambda(2)=\lambda(3)=0$, and represented by the bottom picture of Figure 7.4.

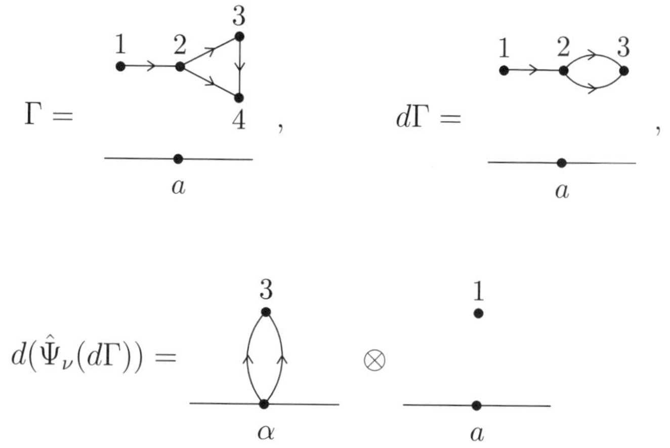

FIGURE 7.4. A diagram Γ for which $d(\widehat{\Psi}_\nu d(\Gamma)) \neq 0$.

Thus

$$d(\widehat{\Psi}_\nu d(\Gamma)) \neq 0.$$

On the other hand

$$d(d\widehat{\Psi}_\nu(\Gamma)) = 0$$

since $d^2 = 0$. Therefore $\widehat{\Psi}_\nu$ is not a chain map.

In fact, the equality $\widehat{\Psi}_\nu d = d\widehat{\Psi}_\nu$ is not really expected to hold. Indeed, if Γ is a diagram with l internal vertices then there are $l^{|P|+1}$ terms in $\Psi_\nu(\Gamma)$, corresponding to the various condensations. On the other hand, for each contractible edge e in Γ, $\Psi_\nu(\Gamma/e)$ has only $(l-1)^{|P|+1}$ terms. Thus there is no clear correspondence between the terms of the sums $\widehat{\Psi}_\nu d(\Gamma)$ and of $d\widehat{\Psi}_\nu(\Gamma)$, and hence no evidence that these two sums should be equal.

This explains why the proof below that Ψ_ν is a chain map is quite elaborate. The idea is to restrict to condensations of Γ for which $\Gamma(\lambda)$ is admissible and to establish Lemma 7.3.8, which amounts to exhibiting a 1–1 correspondence between condensations of Γ and of Γ/e.

Let Γ be a diagram on A.

DEFINITION 7.3.3. A condensation λ of Γ is *admissible* if for each internal vertex i and each $p \in P$ there is an equivalence

$$\lambda(i) = p \quad \Longleftrightarrow \quad i \text{ admits at least two distinct adjacent } p\text{-local vertices.}$$

Denote by $\mathrm{AdmCond}(\Gamma)$ the set of admissible condensations on Γ.

This terminology is motivated by the following:

LEMMA 7.3.4. *If λ is not admissible then $\Gamma(\lambda) \in \mathcal{N}(\nu)$.*

PROOF. Suppose that i is a p-local internal vertex for some $p \in P$, and suppose that it does not have two adjacent p-local vertices. Then i is internal of valence < 2 in $\Gamma(\lambda, p)$, and hence $\Gamma(\lambda, p) \in \mathcal{N}(A_p)$.

Suppose that i is an internal vertex that is not p-local but that has two adjacent p-local vertices for some $p \in P$. Then in $\Gamma(\lambda, 0)$, the external vertex p is connected by a double edge to either i (if $\lambda(i)=0$) or to the external vertex q (if $\lambda(i)=q\in P\setminus\{p\}$). Thus $\Gamma(\lambda,0)\in\mathcal{N}(P)$. □

Lemma 7.3.4 implies that, in $\otimes_{p\in P^*}\mathcal{D}(A_p)$ and for Γ admissible,

$$\Psi_\nu(\Gamma)=\sum_{\lambda\in\mathrm{AdmCond}(\Gamma,\nu)}\Gamma(\lambda). \tag{7.11}$$

LEMMA 7.3.5. *Let λ_1,λ_2 be two admissible condensations on Γ. If λ_1 and λ_2 coincide on all vertices except possibly on one, then $\lambda_1=\lambda_2$.*

PROOF. Let u be a vertex of Γ such that $\lambda_1(v)=\lambda_2(v)$ for $v\neq u$. If u is external then the values of $\lambda_i(u)$ are determined by ν, and hence $\lambda_1=\lambda_2$. Suppose that u is internal. If u has two adjacent vertices that are p-local (for both λ_1 and λ_2) for some $p\in P$, then $\lambda_i(u)=p$ by admissibility. Otherwise $\lambda_i(u)=0$, again by admissibility. □

For a condensation λ of Γ, recall the extension to vertices λ_E from Definition 7.1.2 and $\Gamma(\lambda,p)$ from Equation (7.1).

DEFINITION 7.3.6. An edge e of Γ is *λ-contractible* if it is contractible in $\Gamma(\lambda,\lambda_E(e))$.

LEMMA 7.3.7. *Assume that Γ is admissible and let λ be an admissible condensation.*
An edge e of Γ is λ-contractible *if and only if the following conditions hold:*

(1) e is contractible in Γ, and
(2) $\lambda(s_\Gamma(e))=\lambda(t_\Gamma(e))$ or $\min(\lambda(s_\Gamma(e)),\lambda(t_\Gamma(e)))=0$.

Condition (2) of the lemma is equivalent to having either both endpoints of e p-local for some $p\in P$ or some endpoint being global.

PROOF. First we show that (1) and (2) are necessary conditions. If (1) does not hold then e is a chord in Γ (because Γ is admissible). Therefore e is also a chord in $\Gamma(\lambda,\lambda_E(e))$, and hence it is not λ-contractible. If (2) does not hold then e is a chord in $\Gamma(\lambda,0)$ joining the external vertices $\lambda(s_\Gamma(e))$ and $\lambda(t_\Gamma(e))$. Therefore e is not λ-contractible.

Suppose now that (1) and (2) hold. We will prove that e is a contractible edge in $\Gamma(\lambda,\lambda_E(e))$. Let v and w be the endpoints of e. As Γ is admissible, e is not a loop and $v\neq w$. We distinguish three cases.

- Case 1: $\lambda(v)=\lambda(w)=p\in P$.
  Since e is contractible, v or w is internal in Γ, and hence the same is true in $\Gamma(\lambda,p)$. Clearly e is neither a chord, nor a loop there. Since λ is admissible, if v is internal, then it has an adjacent p-local vertex, distinct from w. Therefore v also has another adjacent vertex in $\Gamma(\lambda,p)$. The same is true for w if it is internal. Therefore e is not a dead end in $\Gamma(\lambda,p)$. Thus e is λ-contractible.
- Case 2: $\lambda(v)=\lambda(w)=0$.
  Then e is an edge of $\Gamma(\lambda,0)$ joining two distinct internal vertices v and w. In particular e is neither a chord nor a loop. Since e is not a dead end in Γ, there exists a vertex $v'\neq w$ that is adjacent to v in Γ. If v' is global then it is also a vertex of $\Gamma(\lambda,0)$ and if v' is p-local, for some $p\in P$, then it becomes an external vertex p in $\Gamma(\lambda,0)$. In both cases in $\Gamma(\lambda,0)$, v has

an adjacent vertex distinct from w. Similarly w has an adjacent vertex in $\Gamma(\lambda, 0)$ distinct from v. Therefore e is not a dead end in $\Gamma(\lambda, 0)$ and e is λ-contractible.

- Case 3: $\lambda(v) = 0$ and $\lambda(w) = p \in P$ (or the other way).
  Then e is an edge of $\Gamma(\lambda, 0)$ joining the internal vertex v to the external vertex p. Since e is not a dead end in Γ, there is a vertex v' adjacent to v in Γ and distinct from w. Since λ is admissible, we have that $\lambda(v') \neq p$, and hence v' is either global, or q-local for some $q \neq p$. Then, in $\Gamma(\lambda, 0)$, v' either is an internal vertex or becomes the external vertex q. In both cases it is a vertex distinct from p and adjacent to v. This proves that e is not a dead end. Thus e is λ-contractible.

□

Let e be a contractible edge in Γ. Let v and w be the endpoints of e with $v < w$. Thus $V_{\Gamma/e} = V_\Gamma \setminus \{w\}$. Define the function

$$\lambda/e \colon V_{\Gamma/e} \longrightarrow P^* \tag{7.12}$$

by

$$(\lambda/e)(z) = \begin{cases} \lambda(z), & \text{if } z \neq v \text{ or } z = v \text{ is external;} \\ \max(\lambda(v), \lambda(w)), & \text{if } z = v \text{ is internal.} \end{cases}$$

It is clear that λ/e is a condensation of Γ/e. Notice also that if e is λ-contractible then $(\lambda/e)(v) = \max(\lambda(v), \lambda(w))$.

Assume that Γ is an admissible diagram. Consider the sets

$$\Omega = \Big\{(e, \lambda) : e \in E_\Gamma,\ \lambda \in \mathrm{AdmCond}(\Gamma),\ e\ \lambda\text{-contractible},$$
$$(\Gamma/e)(\lambda/e) \neq 0 \text{ in } \bigotimes_{p \in P^*} \mathcal{D}(A_p)\Big\},$$
$$\overline{\Omega} = \Big\{(e, \bar\lambda) : e \in E_\Gamma,\ e \text{ contractible},\ \bar\lambda \in \mathrm{Cond}(\Gamma/e),$$
$$(\Gamma/e)(\bar\lambda) \neq 0 \text{ in } \otimes_{p \in P^*} \mathcal{D}(A_p)\Big\},$$

and the map

$$\omega \colon \Omega \longrightarrow \overline{\Omega}$$
$$(e, \lambda) \longmapsto (e, \lambda/e).$$

LEMMA 7.3.8. *ω is a bijection.*

PROOF. We first show that ω is injective. Let e be a contractible edge and, for $i = 1, 2$, let λ_i be admissible condensations of Γ such that e is λ_i-contractible and $(\Gamma/e)(\lambda_i/e) \neq 0$. Assume that $\lambda_1/e = \lambda_2/e$. We will show that $\lambda_1 = \lambda_2$.

Set $\bar\lambda = \lambda_1/e = \lambda_2/e$. This is an admissible condensation because $(\Gamma/e)(\bar\lambda) \neq 0$ in $\otimes_{p \in P^*} \mathcal{D}(A_p)$ and because of Lemma 7.3.4. Let v and w be the endpoints of e with $v < w$. Thus $V_{\Gamma/e} = V_\Gamma \setminus \{w\}$. We know that λ_i agrees with $\bar\lambda$ on $V_\Gamma \setminus \{v, w\}$, and therefore we only need to show that $\lambda_1(v) = \lambda_2(v)$ and $\lambda_1(w) = \lambda_2(w)$. Moreover, since each λ_i is admissible, by Lemma 7.3.5 it is enough to prove only one of these two equations.

If v is external, then $\lambda_1(v) = \lambda_2(v)$ is determined and hence $\lambda_1 = \lambda_2$. Suppose that v is internal. If $\bar\lambda(v) = 0$ then, since $\bar\lambda(v) = \max(\lambda_i(v), \lambda_i(w))$, we get $\lambda_i(v) = \lambda_i(w) = 0$ for $i = 1, 2$, and hence $\lambda_1 = \lambda_2$. Suppose that $\bar\lambda(v) = p \in P$. By admissibility of $\bar{\lambda}$, there exist two vertices x and y other than v and w that are adjacent to v in Γ/e and with $\bar{\lambda}(x) = \bar{\lambda}(y) = p$. This implies that $\lambda_i(x) = \lambda_i(y) = p$ for $i = 1, 2$. Then in Γ, either x and y are both adjacent to v (respectively to w), or x is adjacent to v and y is adjacent to w (or the other way around). In the first case we get by admissibility of λ_i that $\lambda_1(v) = \lambda_2(v) = p$ (respectively $\lambda_1(w) = \lambda_2(w) = p$), and hence $\lambda_1 = \lambda_2$ by Lemma 7.3.5. In the second case, since $p = \bar{\lambda}(v) = \max(\lambda_1(v), \lambda_1(w))$, we get that $\lambda_1(v) = p$ or $\lambda_1(w) = p$. Let us say that $\lambda_1(v) = p$, the other case being analogous. Then w is adjacent to v and to either x or y, and thus w is adjacent to two p-local vertices (for the condensation λ_1), and hence we also have $\lambda_1(w) = p$ by admissibility. The same argument shows that $\lambda_2(v) = \lambda_2(w) = p$. This proves injectivity of ω.

To show ω is surjective, let e be a contractible edge of Γ and let $\bar{\lambda}$ be a condensation of Γ/e such that $(\Gamma/e)(\bar{\lambda}) \neq 0$. We will construct an admissible condensation λ of Γ such that e is λ-contractible and $\lambda/e = \bar{\lambda}$. Let v and w again be the endpoints of e with $v < w$. For $z \in V_\Gamma \setminus \{v, w\}$, set $\lambda(z) = \bar{\lambda}(z)$. We need to define $\lambda(v)$ and $\lambda(w)$ and to check that λ has the desired properties. We consider the following cases.

(1) Suppose v is external. Then $\lambda(v) = \bar{\lambda}(v) = p \in P$ is prescribed by ν.
   (a) Suppose that there exists a vertex x in Γ different from v and adjacent to w such that $\bar{\lambda}(x) = p$. In that case, set $\lambda(v) = \lambda(w) = p$. Then λ is admissible at the vertex w because it has two p-local adjacent vertices v and x. It is easy to check that λ is also admissible at the other internal vertices of Γ, using the fact that $\bar{\lambda}$ is. Moreover e is λ-contractible by Lemma 7.3.7 since it is contractible and $\lambda(v) = \lambda(w)$.
   (b) Suppose that w is not adjacent to any p-local vertex other than v. In that case, set $\lambda(v) = p$ and $\lambda(w) = 0$. The vertex w is not adjacent in Γ to two vertices x and y such that $\bar{\lambda}(x) = \bar{\lambda}(y) = q \in P$ with $q \neq p$ because otherwise $(\Gamma/e)(\bar{\lambda}, 0)$ would contain a double edge joining p and q, and hence $\Gamma/e \in \mathcal{N}(\nu)$, contrary to our hypothesis. This proves that λ is admissible at w and the admissibility at other vertices is a consequence of the admissibility of $\bar{\lambda}$. Also e is λ-contractible.

(2) Suppose that v is internal. Then w is also internal since $v < w$.
   (a) Suppose that $\bar{\lambda}(v) = 0$. In that case set $\lambda(v) = \lambda(w) = 0$. By admissibility of $\bar{\lambda}$, there do not exist two vertices $x, y \in V_\Gamma \setminus \{v, w\}$ adjacent in Γ to either v or w with $\bar{\lambda}(x) = \bar{\lambda}(y) \in P$. It is easy to see that λ is admissible and e is λ-contractible.
   (b) Suppose that $\bar{\lambda}(v) = p \in P$. By admissibility of $\bar{\lambda}$ there exist two distinct vertices $x, y \in V_\Gamma \setminus \{v, w\}$ adjacent in Γ to either v or w such that $\bar{\lambda}(x) = \bar{\lambda}(y) = p$.
      - If v is not adjacent to any vertices in $(V_\Gamma \setminus \{v, w\}) \cap \bar{\lambda}^{-1}(p)$ then set $\lambda(v) = 0$ and $\lambda(w) = p$.
      - If w is not adjacent to any vertices in $(V_\Gamma \setminus \{v, w\}) \cap \bar{\lambda}^{-1}(p)$ then set $\lambda(v) = p$ and $\lambda(w) = 0$.
      - If both v and w are adjacent to some vertices in $(V_\Gamma \setminus \{v, w\}) \cap \bar{\lambda}^{-1}(p)$ then set $\lambda(v) = \lambda(w) = p$.

   In each case it is easy to see that λ is admissible and that e is λ-contractible.

This proves surjectivity of ω. $\square$

LEMMA 7.3.9. *If Γ is admissible, if λ is admissible, and if e is a λ-contractible edge of Γ, then, for $p \in P^*$, we have in $\mathcal{D}(A_p)$:*

$$(\Gamma/e)(\lambda/e, p) = \begin{cases} \Gamma(\lambda, p)/e, & \text{if } p = \lambda_E(e); \\ \Gamma(\lambda, p), & \text{otherwise.} \end{cases}$$

PROOF. Let v and w be the endpoints of e with $v < w$. Then $V_{\Gamma/e} = V_\Gamma \setminus \{w\}$ and $E_{\Gamma/e} = E_\Gamma \setminus \{e\}$. It is easy to see that the equations to prove are equivalent to

$$\begin{cases} \lambda/e &= \lambda|(V_\Gamma \setminus \{w\}) \\ (\lambda/e)_E &= \lambda_E|(E_\Gamma \setminus \{e\}) \end{cases}$$

Since e is λ-contractible, by Lemma 7.3.7, $\lambda(v) = \lambda(w)$ or $\min(\lambda(v), \lambda(w)) = 0$. If $\lambda(v) < \lambda(w)$ then $\lambda(v) = 0$ which implies that v is internal, and the same for w because $v < w$, in which case we can transpose the order of v and w to get an equivalent diagram (up to sign) in which the roles of v and w are exchanged. Therefore, without loss of generality we can always assume that $\lambda(v) \geq \lambda(w)$. This implies that $(\lambda/e)(v) = \lambda(v)$. Also for $z \neq v, w$ we have $(\lambda/e)(z) = \lambda(z)$. Thus $\lambda/e = \lambda|V_\Gamma \setminus \{w\}$.

It remains to prove that $(\lambda/e)_E = \lambda_E|(E_\Gamma \setminus \{e\})$. Let $f \neq e$ be an edge of Γ. If w is not an endpoint of f, then, since $\lambda/e = \lambda|V_\Gamma \setminus \{w\}$, $(\lambda/e)_E(f) = \lambda_E(f)$. Suppose that w is an endpoint of f. If $\lambda(w) = \lambda(v)$ then $(\lambda/e)_E(f) = \lambda_E(f)$. Otherwise $\lambda(w) = 0$ and $\lambda(v) = p \in P$, and hence f is global in Γ. As Γ is admissible and $f \neq e$, the other endpoint of f is not v. Since λ is admissible and since w is not r-local but is adjacent to the p-local vertex v, we get that the other endpoint of f is not p-local. This implies that f is global in Γ/e, and hence $(\lambda/e)_E(f) = \lambda_E(f) = 0$. This proves that $(\lambda/e)_E = \lambda_E|(E_\Gamma \setminus \{e\})$. $\square$

PROOF OF PROPOSITION 7.3.1. Let Γ be an admissible diagram on A. For $p \in P^*$ and for a condensation λ of Γ, define the sign

$$\eta(\Gamma, \lambda, p) := (-1)^s \quad \text{with} \quad s = \sum_{\substack{q \in P^* \\ q < p}} \deg(\Gamma(\lambda, q)).$$

We have

$$d(\Psi_\nu(\Gamma)) \overset{(7.11)}{=} d\left(\sum_{\lambda\in\mathrm{AdmCond}(\Gamma)} \Gamma(\lambda)\right)$$

$$= \sum_{\lambda\in\mathrm{AdmCond}(\Gamma)} \sum_{p\in P^*} \sigma(\Gamma,\lambda)\cdot\eta(\Gamma,\lambda,p)\cdot \tag{7.13}$$

$$\cdot \bigotimes_{q<p} \Gamma(\lambda,q) \otimes d\left(\Gamma(\lambda,p)\right) \otimes \bigotimes_{q>p} \Gamma(\lambda,q)$$

$$= \sum_{\lambda\in\mathrm{AdmCond}(\Gamma)} \sum_{p\in P^*} \sum_{e\in E^{\mathrm{contr}}_{\Gamma(\lambda,p)}} \sigma(\Gamma,\lambda)\cdot\eta(\Gamma,\lambda,p)\cdot\epsilon(\Gamma(\lambda,p),e)\cdot$$

$$\cdot \bigotimes_{q<p} \Gamma(\lambda,q) \otimes \Gamma(\lambda,p)/e \otimes \bigotimes_{q>p} \Gamma(\lambda,q)$$

$$\overset{\text{Lemma 7.3.9}}{=} \sum_{(e,\lambda)\in\Omega} \sigma(\Gamma,\lambda)\cdot\eta(\Gamma,\lambda,\lambda_E(e))\cdot\epsilon(\Gamma(\lambda,\lambda_E(e)),e)\cdot \tag{7.14}$$

$$\cdot\left(\bigotimes_{p\in P^*} (\Gamma/e)(\lambda/e,p)\right). \tag{7.15}$$

On the other hand,

$$\Psi_\nu(d(\Gamma)) = \Psi_\nu\left(\sum_{e\in E^{\mathrm{contr}}_\Gamma} \epsilon(\Gamma,e)\cdot\Gamma/e\right)$$

$$= \sum_{(e,\bar\lambda)\in\overline{\Omega}} \epsilon(\Gamma,e)\cdot\sigma(\Gamma/e,\bar\lambda)\cdot\left(\bigotimes_{p\in P^*}(\Gamma/e)(\bar\lambda,p)\right)$$

$$\overset{\text{Lemma 7.3.8}}{=} \sum_{(e,\lambda)\in\Omega} \epsilon(\Gamma,e)\cdot\sigma(\Gamma/e,\lambda/e)\cdot\left(\bigotimes_{p\in P^*}(\Gamma/e)(\lambda/e,p)\right) \tag{7.16}$$

It remains to check that the signs of (7.15) and (7.16) agree, which is straighforward. □

7.4. Proof that the cooperad structure is well-defined

We show in this section that $\{\widehat{\Psi}_\nu\}$ and $\{\Psi_\nu\}$ endow $\widehat{\mathcal{D}}$ and $\mathcal{D}$ with the cooperadic structure (the former in the category of vector spaces and the latter in the category of chain complexes), when ν runs over all weak ordered partitions.

First let us show the associativity of the structure maps. Suppose given a weak ordered partition $\nu\colon A\to P$ as before. Suppose moreover that A is itself linearly ordered, that ν is increasing, and that $P^*\cap A=\emptyset$. Let $\xi\colon B\to A$ be an ordered weak partition of a finite set B. Set $B_a=\xi^{-1}(a)$ for $a\in A$. Also set $A^*=\{0\}\otimes A$.

We then have a natural bijection

$$\coprod_{a\in A} B_a \cong \coprod_{p\in P}\coprod_{a\in A_p} B_a.$$

For $p\in P$, the partition ξ restricts to a weak ordered partition

$$\xi_p\colon \coprod_{a\in A_p} B_a \longrightarrow A_p.$$

The associativity of $\widehat{\Psi}$ amounts to the following lemma whose proof is straightforward.

LEMMA 7.4.1. *The following diagram is commutative:*

(7.17)

$$\begin{array}{ccc}
\widehat{\mathcal{D}}\Big(\coprod_{p\in P}\Big(\coprod_{a\in A_p} B_a\Big)\Big) & =\!=\!=\!=\!=\!= & \widehat{\mathcal{D}}\Big(\coprod_{a\in A} B_a\Big) \\
\Big\downarrow {\scriptstyle \widehat{\Psi}_{\nu\circ\xi}} & & \Big\downarrow {\scriptstyle \widehat{\Psi}_{\xi}} \\
\widehat{\mathcal{D}}(P)\otimes \bigotimes_{p\in P}\widehat{\mathcal{D}}\Big(\coprod_{a\in A_p} B_a\Big) & & \widehat{\mathcal{D}}(A)\otimes \bigotimes_{a\in A}\widehat{\mathcal{D}}(B_a) \\
\Big\downarrow {\scriptstyle \mathrm{id}\otimes \bigotimes_{p\in P}\widehat{\Psi}_{\xi_p}} & & \Big\downarrow {\scriptstyle \widehat{\Psi}_{\nu}\otimes\mathrm{id}} \\
\widehat{\mathcal{D}}(P)\otimes \bigotimes_{p\in P}\Big(\widehat{\mathcal{D}}(A_p)\otimes\bigotimes_{a\in A_p}\widehat{\mathcal{D}}(B_a)\Big) & \xrightarrow[\tau]{\cong} & \Big(\widehat{\mathcal{D}}(P)\otimes\bigotimes_{p\in P}\widehat{\mathcal{D}}(A_p)\Big)\otimes\bigotimes_{a\in A}\widehat{\mathcal{D}}(B_a)
\end{array}$$

The horizontal bottom isomorphism τ is the obvious reordering of factors (with the usual Koszul sign).

We next define an action of the group $\mathrm{Perm}(A)$ of permutation of the finite set A on diagrams on A. Given a permutation $\sigma \in \mathrm{Perm}(A)$ and a diagram $\Gamma = (A, E, I, s, t)$, we define a new diagram

$$\sigma \cdot \Gamma = (A, E, I, \sigma \circ s, \sigma \circ t)$$

where the bijection $\sigma\colon A \xrightarrow{\cong} A$ is extended to all vertices by $\sigma(i) = i$ for $i \in I$. The following is immediate.

PROPOSITION 7.4.2. *There is an induced action of ($\mathbb{Z}$-graded) CDGA of* $\mathrm{Perm}(A)$ *on $\widehat{\mathcal{D}}(A)$ and $\mathcal{D}(A)$.*

To define the counits of the cooperad structure, consider the CDGA maps

$$\hat{\eta}\colon \widehat{\mathcal{D}}(1) \longrightarrow \mathbb{K} \quad \text{and} \quad \eta\colon \mathcal{D}(1) \longrightarrow \mathbb{K}$$

defined by $\hat{\eta}(\mathbf{1}) = 1$ and $\hat{\eta}(\Gamma) = 0$ for a diagram other than the unit, and similarly for η.

THEOREM 7.4.3. *The structure maps $\widehat{\Psi}_\nu$ and Ψ_ν, for all weak ordered partitions ν, the symmetric action, and the counits $\hat{\eta}$ and η described above define:*

- *the structure of a cooperad of $\mathbb{Z}$-graded $\mathbb{K}$-algebras on $\widehat{\mathcal{D}}$, and*
- *the structure of a cooperad of CDGAs on $\mathcal{D}$ ($\mathbb{Z}$-graded if $N = 2$).*

PROOF. The associativity of the structure maps $\widehat{\Psi}$ required for a cooperad structure is exactly Lemma 7.4.1. We have the corresponding associativity for Ψ since, by Proposition 7.1.5, that structure map is induced by $\widehat{\Psi}$. It is easy to check that $\hat{\eta}$ and η are counits. The equivariance is also easy to check. $\square$

Note that the cooperad structures developed here are related to cooperad strucures on the category of sets (as developed in [**28**]).

CHAPTER 8

Equivalence of the cooperads $\mathcal{D}$ and $\mathrm{H}^*(\mathrm{C}[\bullet])$

We show in this chapter that the CDGA cooperad $\mathcal{D}$ of admissible diagrams is weakly equivalent to the cohomology algebra of the Fulton-MacPherson cooperad $\mathrm{H}^*(\mathrm{C}[\bullet];\mathbb{K})$ for any commutative ring with unit $\mathbb{K}$ and ambient dimension $N \geq 2$.

Fix a finite set A. We first recall the computation of the algebra $\mathrm{H}^*(\mathrm{C}[A];\mathbb{K})$ due to F. Cohen [**9**]. Denote by $[\mathrm{vol}] \in \mathrm{H}^{N-1}(S^{N-1};\mathbb{K})$ the orientation class of the sphere. For a, b which are distinct in A, recall the map $\theta_{ab}\colon \mathrm{C}[A] \to S^{N-1}$ from (5.6) which gives the direction between two points of the configuration, and set

$$g_{ab} := \theta_{ab}^*([\mathrm{vol}]) \in \mathrm{H}^{N-1}(\mathrm{C}[A];\mathbb{K}). \tag{8.1}$$

Then as graded algebras we have

$$\mathrm{H}^*(\mathrm{C}[A];\mathbb{K}) = \frac{\wedge(\{g_{ab} : a, b \in A,\ a \neq b\})}{(\text{3-term relation}\,;\ (g_{ab})^2\,;\ g_{ab} - (-1)^N g_{ba})}$$

where $\wedge(\{g_{ab}\})$ is the free commutative graded $\mathbb{K}$-algebra generated by the g_{ab}'s, and the 3-*term relation* is

$$g_{ab}g_{bc} + g_{bc}g_{ca} + g_{ca}g_{ab}$$

for all distinct $a, b, c \in A$. Here we follow the standard conventions in rational homotopy theory and denote by $\wedge Z$ the free commutative graded algebra generated by a graded vector space Z. This is thus the tensor product of the symmetric algebra on Z^{even} and the exterior algebra on Z^{odd}.

For a, b distinct in A, denote by

$$\Gamma\langle a, b\rangle \tag{8.2}$$

the diagram on A with no internal vertices and whose only edge is a chord from a to b. This is an admissible cocycle of degree $N-1$.

We endow the cohomology algebra with a zero differential to make it a CDGA.

THEOREM 8.1. *For $N \geq 2$, there is a quasi-isomorphism of CDGAs ($\mathbb{Z}$-graded if $N = 2$)*

$$\bar{\mathrm{I}}\colon \mathcal{D}(A) \xrightarrow{\simeq} (\mathrm{H}^*(\mathrm{C}[A];\mathbb{K}), 0)$$

characterized by

$$\begin{cases} \bar{\mathrm{I}}(\Gamma\langle a, b\rangle) = g_{ab}, & \text{for } a, b \text{ distinct in } A;\\ \bar{\mathrm{I}}(\Gamma) = 0, & \text{for a diagram } \Gamma \text{ with internal vertices.}\end{cases}$$

Moreover $\bar{\mathrm{I}}$ is a weak equivalence of cooperads.

The rest of the chapter is devoted to the proof of this theorem.

Consider the submodule $\mathcal{D}^{(0)}(A)$ of $\mathcal{D}(A)$ generated by admissible diagrams without internal vertices. Then

$$\mathcal{D}^{(0)}(A) = \frac{\wedge(\{\Gamma\langle a,b\rangle : a,b \in A,\, a \neq b\})}{((\Gamma\langle a,b\rangle)^2\,;\,\Gamma\langle a,b\rangle - (-1)^N\Gamma\langle b,a\rangle)}.$$

Therefore we have a surjective algebra map

$$\bar{\mathrm{I}}_0\colon \mathcal{D}^{(0)}(A) \longrightarrow \mathrm{H}^*(\mathrm{C}[A];\mathbb{K})$$

defined by $\bar{\mathrm{I}}_0(\Gamma\langle a,b\rangle) = g_{ab}$.

LEMMA 8.2.

$$\bar{\mathrm{I}}_0(\mathcal{D}^{(0)}(A) \cap d(\mathcal{D}(A))) = 0.$$

PROOF. It is enough to prove that $\bar{\mathrm{I}}_0(d\Gamma) = 0$ when Γ is an admissible diagram consisting of one internal vertex i and n edges connecting it to the external vertices $a_1,\ldots,a_n$. In that case,

$$\bar{\mathrm{I}}_0(d\Gamma) = \sum_{k=1}^{n}(-1)^k g_{a_1a_k}g_{a_2a_k}\cdots g_{a_{k-1}a_k}g_{a_ka_{k+1}}\cdots g_{a_ka_n}.$$

The right side vanishes as can be seen using an easy induction on $n \geq 3$ and the 3-term relation

$$g_{a_1a_k}g_{a_2a_k} = g_{a_1a_2}(g_{a_2a_k} - g_{a_1a_k}).$$

from (1.3) (which corresponds to the case $n = 3$).

When $\mathbb{K} = \mathbb{R}$, an alternative non-computational proof is possible: The Kontsevich configuration space integral

$$\mathrm{I}\colon \mathcal{D}(n) \to \Omega_{PA}(\mathrm{C}[n])$$

(to be defined in Chapter 9) commutes with the differential (Proposition 9.4.1), and hence

$$\bar{\mathrm{I}}_0(d\Gamma) = [\mathrm{I}(d\Gamma)] = [d\mathrm{I}(\Gamma)] = 0$$

in $\mathrm{H}(\Omega_{PA}(\mathrm{C}[n])) \cong \mathrm{H}^*(\mathrm{C}[n];\mathbb{R})$. □

This lemma implies that we can define the CDGA morphism $\bar{\mathrm{I}}$ by

$$\bar{\mathrm{I}}(\Gamma) = \begin{cases} \bar{\mathrm{I}}_0(\Gamma), & \text{if } \Gamma \text{ has no internal vertices;} \\ 0, & \text{otherwise.} \end{cases} \tag{8.3}$$

It is straightforward to check that this induces a morphism of cooperads.

Since $\bar{\mathrm{I}}$ induces a surjection in homology, in order to prove that it is a quasi-isomorphism we only need to establish the following

LEMMA 8.3. *The graded* $\mathbb{K}$*-modules* $\mathrm{H}_*(\mathcal{D}(A))$ *and* $\mathrm{H}_*(\mathrm{C}[A];\mathbb{K})$ *are isomorphic.*

The proof of this lemma will take up the rest of this chapter.

A diagram Γ on A induces a partition of A into its path-connected components, and we denote this partition by ν_Γ. In other words, two external vertices a and b belong to the same element $C \in \nu_\Gamma$ (see Definition 2.3.1 for definitions regarding partitions) if and only if they are connected by a path of unoriented edges in Γ. For a partition ν of A, denote by

$$\mathcal{D}(A)\langle\nu\rangle$$

the submodule of $\mathcal{D}(A)$ generated by admissible diagrams Γ whose partition of connected components is ν. It is clear that $\mathcal{D}(A)\langle\nu\rangle$ is a subcomplex of $\mathcal{D}(A)$. In the particular case of the indiscrete partition $\nu = \{A\}$, we get the subcomplex of *connected admissible diagrams*

$$\underset{\sim}{\mathcal{D}}(A) := \mathcal{D}(A)\langle\{A\}\rangle.$$

We have an isomorphism of complexes

$$\mathcal{D}(A) \cong \bigoplus_{\nu} \bigotimes_{C \in \nu} \underset{\sim}{\mathcal{D}}(C) \tag{8.4}$$

where the sum runs over all partitions ν of the set A.

The Poincaré series of the homology of the configuration space $\mathrm{C}[A]$ is given by [**9**]

$$(1+t)(1+2t)\dots(1+(|A|-1)t) \tag{8.5}$$

with t of degree $N-1$. In particular the top degree Betti number is

$$\dim \mathrm{H}^{(N-1)(|A|-1)}(\mathrm{C}(A); \mathbb{K}) = (|A|-1)! \tag{8.6}$$

In view of the isomorphism (8.4) and formulas (8.5) and (8.6), Lemma 8.3 will be a direct consequence of the following

LEMMA 8.4. *For A non-empty,*

$$\dim \mathrm{H}^i(\underset{\sim}{\mathcal{D}}(A)) = \begin{cases} (|A|-1)!, & \textit{if } i = (N-1)\cdot(|A|-1), \\ 0, & \textit{otherwise.} \end{cases}$$

Before proving this lemma, we introduce further submodules. Fix an element $a \in A$ and consider the following submodules of $\underset{\sim}{\mathcal{D}}(A)$:

- $\mathcal{U}_0$ is the submodule generated by connected admissible diagrams with a of valence 1 and such that the only edge with endpoint a is contractible;
- $\mathcal{U}_1$ is the submodule generated by connected admissible diagrams with a of valence ≥ 2;
- $\underset{\sim}{\mathcal{D}}'(A)$ is the submodule generated by all connected admissible diagrams that are neither in $\mathcal{U}_0$ nor in $\mathcal{U}_1$.

It is clear that $\underset{\sim}{\mathcal{D}}'(A)$ is a subcomplex of $\underset{\sim}{\mathcal{D}}(A)$.

LEMMA 8.5. *The inclusion*

$$\underset{\sim}{\mathcal{D}}'(A) \hookrightarrow \underset{\sim}{\mathcal{D}}(A)$$

is a quasi-isomorphism.

PROOF. Consider the quotient complex $\mathcal{U} := \underset{\sim}{\mathcal{D}}(A)/\underset{\sim}{\mathcal{D}}'(A)$. We need to show that $\mathcal{U}$ is acyclic.

Identify $\mathcal{U}$ in the obvious way with the graded $\mathbb{K}$-module $\mathcal{U}_0 \oplus \mathcal{U}_1$, and define an increasing filtration on $\mathcal{U}$ where elements of filtration $\leq p$ are the linear combinations of diagrams in $\mathcal{U}_0$ with less than p edges and diagrams in $\mathcal{U}_1$ with less than $p-1$ edges. The differential preserves the filtration. Consider the spectral sequence associated to this filtration and which converges to the homology of $\mathcal{U}$. The differential at the 0th page

$$d^0 : \mathcal{U}_0 \longrightarrow \mathcal{U}_1$$

consists of contracting the only edge with endpoint a. It is an isomorphism because there is an inverse given by "blowing up" the vertex a of a diagram $\Gamma \in \mathcal{U}_1$ into a contractible edge (a, a') as in Figure 8.1. Therefore the page E^1 of the spectral sequence is trivial and hence $\mathcal{U}$ is acyclic. □

FIGURE 8.1. Example of blowing up vertex $a = 1$ into a contractible edge (a, a').

We are now ready for the

PROOF OF LEMMA 8.4. The proof is by induction on the cardinality of A.

If A is a singleton then $\underset{\sim}{\mathcal{D}}'(A) = \mathbb{K} \cdot \mathbf{1}$, where $\mathbf{1}$ is the unit diagram with a single external vertex and no internal vertices or edges. Lemma 8.4 is then a consequence of Lemma 8.5.

Let A be of cardinality $k \geq 2$ and suppose that the lemma has been proved for $< k$ external vertices. Fix $a \in A$. Any diagram in $\underset{\sim}{\mathcal{D}}'(A)$ has exactly one edge with endpoint a and it is a chord. We have an isomorphism of complexes

$$\underset{\sim}{\mathcal{D}}'(A) \cong \bigoplus_{b \in A \setminus \{a\}} \Gamma\langle a, b\rangle \cdot \underset{\sim}{\mathcal{D}}(A \setminus \{a\}).$$

Using Lemma 8.5 we conclude that

$$\dim \mathrm{H}^i(\underset{\sim}{\mathcal{D}}(A)) = (|A| - 1) \cdot \dim \mathrm{H}^{i-(N-1)}(\underset{\sim}{\mathcal{D}}(A \setminus \{a\}))$$

and deduce the desired conclusion using the induction hypothesis. □

We now finish the

PROOF OF LEMMA 8.3. An elementary computation by induction on $|A|$ using isomorphism (8.4) and Lemma 8.4 shows that the Poincaré series of $\mathrm{H}^*(\mathcal{D}(A))$ is exactly (8.5), and this is also the Poincaré series of $\mathrm{H}^*(\mathrm{C}[A])$. □

This finishes the proof of Theorem 8.1.

CHAPTER 9

The Kontsevich configuration space integrals

In the previous chapter we built a quasi-isomorphism

$$\bar{\mathrm{I}}\colon \mathcal{D}(n) \xrightarrow{\simeq} \mathrm{H}^*(\mathrm{C}[n])$$

of cooperads. The goal of this section is to construct a CDGA morphism

$$\mathrm{I}\colon \mathcal{D}(n) \longrightarrow \Omega_{PA}(\mathrm{C}[n])$$

which will turn out to be a quasi-isomorphism as well as "almost" a morphism of cooperads (see Proposition 9.5.1 below for the precise meaning of this.) In this entire chapter the ground ring is the field of real numbers $\mathbb{K} = \mathbb{R}$. We also fix an integer $N \geq 2$ which is the dimension of the euclidean space $\mathbb{R}^N$ on which we consider the configuration spaces $\mathrm{C}[n]$, as well as the underlying dimension of the space of admissible diagrams $\mathcal{D} = \mathcal{D}_N$.

We will throughout use many constructions related to semi-algebraic forms that we quickly reviewed in Chapter 4 and which are fully developed in [**18**].

The plan of this chapter is as follows.

9.1: We construct a linear map

$$\widehat{\mathrm{I}}\colon \widehat{\mathcal{D}}(n) \longrightarrow \Omega_{PA}(\mathrm{C}[n]).$$

9.2: We prove that $\widehat{\mathrm{I}}$ is a map of algebras.

9.3: We show that $\widehat{\mathrm{I}}$ induces the desired map I on $\mathcal{D}(n)$ by showing that it vanishes on non-admissible diagrams.

9.4: We prove that $\widehat{\mathrm{I}}$, and hence I, commutes with the differentials.

9.5: We prove that $\widehat{\mathrm{I}}$ and I are almost morphisms of cooperads.

Before reading on, the reader might want to look at the last part of the introduction where some intuition about the integrals defined in this chapter is given.

9.1. Construction of the Kontsevich configuration space integral $\hat{\mathrm{I}}$

Fix a finite set A. We construct a linear map

$$\widehat{\mathrm{I}}\colon \widehat{\mathcal{D}}(A) \longrightarrow \Omega_{PA}(\mathrm{C}[A])$$

as follows.

Let Γ be a diagram on A. Let vol be the standard normalized volume form on the sphere $S^{N-1} \subset \mathbb{R}^N$ defined as

$$\mathrm{vol} = \kappa_N \cdot \sum_{i=1}^{N} (-1)^i t_i \, dt_1 \wedge \cdots \wedge \widehat{dt_i} \wedge \cdots \wedge dt_N \tag{9.1}$$

where $t_1, \ldots, t_N$ are the standard coordinates in $\mathbb{R}^N$, $\widehat{dt_i}$ means dt_i is omitted, and $\kappa_N \in \mathbb{R}$ is a normalizing constant such that

$$\int_{S^{N-1}} \mathrm{vol} = 1.$$

Since all the functions in (9.1) are polynomials, and are hence semi-algebraic, $\mathrm{vol} \in \Omega^{N-1}_{\min}(S^{N-1})$ is what was called in Chapter 4 a *minimal form*. More generally, for any linearly ordered finite set E, consider the product of spheres

$$(S^{N-1})^E = \prod_{e\in E} S^{N-1},$$

and denote by vol_E the top volume form in that product, that is,

$$\mathrm{vol}_E := \times_{e\in E}\, \mathrm{vol}_e \in \Omega_{\min}((S^{N-1})^E) \tag{9.2}$$

where the products are taken in the order of E and vol_e is the standard normalized volume form on the eth factor.

For v and w two distinct vertices in V_Γ, recall from (5.6) the map

$$\theta_{v,w}\colon \mathrm{C}[V_\Gamma] \longrightarrow S^{N-1}$$

which associates to a configuration x the direction from $x(v)$ to $x(w)$. By convention, when $v = w$, we set $\theta_{v,v}$ to be the constant map to a fixed basepoint of the sphere. For an edge e of Γ we set $\theta_e = \theta_{s_\Gamma(e),t_\Gamma(e)}$ and we define

$$\theta_\Gamma := (\theta_e)_{e\in E_\Gamma} : \mathrm{C}[V_\Gamma] \longrightarrow (S^{N-1})^{E_\Gamma}.$$

Recall the definition of a minimal form from Equation (4.3). We then have such a form

$$\theta^*_\Gamma(\mathrm{vol}_{E_\Gamma}) \in \Omega_{\min}(\mathrm{C}[V_\Gamma]) \tag{9.3}$$

which is of degree $l = |E_\Gamma| \cdot (N-1)$.

By Theorem 5.3.2, the canonical projection

$$\pi_\Gamma\colon \mathrm{C}[V_\Gamma] \longrightarrow \mathrm{C}[A] \tag{9.4}$$

is an oriented SA bundle. When $|A| \geq 2$, the fiber of π_Γ is of dimension $N \cdot |I_\Gamma|$ and integration along the fiber [**18**, Definition 8.3] gives a pushforward map

$$(\pi_\Gamma)_*\colon \Omega^l_{\min}(\mathrm{C}[V_\Gamma]) \longrightarrow \Omega^{l-N\cdot|I_\Gamma|}_{PA}(\mathrm{C}[A]). \tag{9.5}$$

When $|A| \geq 2$, define $\widehat{\mathrm{I}}(\Gamma)$ as the pushfoward

$$\widehat{\mathrm{I}}(\Gamma) := (\pi_\Gamma)_*(\theta^*_\Gamma(\mathrm{vol}_{E_\Gamma})) \in \Omega_{PA}(\mathrm{C}[A]). \tag{9.6}$$

For example when Γ is the diagram from Figure 1.2 in the Introduction, $\widehat{\mathrm{I}}(\Gamma)$ corresponds to formula (1.8).

If A is empty or a singleton we just set

$$\widehat{\mathrm{I}}(\Gamma) := \begin{cases} 1, & \text{if } \Gamma \text{ is the unit diagram;} \\ 0, & \text{otherwise.} \end{cases} \tag{9.7}$$

The reason we treat the case $|A| \leq 1$ separately is that the dimension of the fiber of π_Γ is then smaller than expected when there are internal vertices (see Theorem 5.3.2). Therefore we should in those cases consider the pushforward $\pi_{\Gamma*}$ of (9.5) to be 0. Formula (9.7) is a clean way to do this.

LEMMA 9.1.1. *For any finite set A, formulas (9.6) and (9.7) induce a degree 0 linear map*

$$\widehat{\mathrm{I}}\colon \widehat{\mathcal{D}}(A) \longrightarrow \Omega_{PA}(\mathrm{C}[A]).$$

PROOF. This is clear for $|A| \leq 1$. Suppose that $|A| \geq 2$. It is easy to check that (9.6) is compatible with the equivalence relation $\simeq$ of Definition 6.2.2 (it is the compatibility with $\widehat{\mathrm{I}}$ which is the motivation for the definition of $\simeq$). We extend it by linearity. It is clear that $\widehat{\mathrm{I}}$ is of degree 0 (recall Definition 6.2.3 of the degree of a diagram). □

9.2. $\widehat{\mathrm{I}}$ is a morphism of algebras

In this section we prove

PROPOSITION 9.2.1. *$\widehat{\mathrm{I}}$ is a morphism of algebras.*

PROOF. If $|A| \leq 1$ then the proposition is obvious. Suppose now that $|A| \geq 2$. Let Γ_1 and Γ_2 be two diagrams on A and suppose, without loss of generality, that they have disjoint sets of internal vertices and of edges. Notice that $V_{\Gamma_1 \cdot \Gamma_2} = V_{\Gamma_1} \cup_A V_{\Gamma_2}$ and consider the pullback

$$\begin{array}{ccc} \mathrm{C}^{\mathrm{sing}}[V_{\Gamma_1}, V_{\Gamma_2}] & \xrightarrow{q_2} & \mathrm{C}[V_{\Gamma_2}] \\ {\scriptstyle q_1}\downarrow & \text{pullback} & \downarrow{\scriptstyle \pi_2} \\ \mathrm{C}[V_{\Gamma_1}] & \xrightarrow{\pi_1} & \mathrm{C}[A] \end{array} \tag{9.8}$$

which defines a singular configuration space as in Section 5.5.

Set $\pi' = \pi_i \circ q_i\colon\ \mathrm{C}^{\mathrm{sing}}[V_{\Gamma_1}, V_{\Gamma_2}] \to \mathrm{C}[A]$. Consider the canonical projections

$$\pi\colon\ \mathrm{C}[V_{\Gamma_1 \cdot \Gamma_2}] \longrightarrow \mathrm{C}[A]$$

and

$$\rho_i\colon\ \mathrm{C}[V_{\Gamma_1 \cdot \Gamma_2}] \longrightarrow \mathrm{C}[V_{\Gamma_i}]$$

for $i = 1, 2$, and the induced map to the pullback

$$\rho\colon\ \mathrm{C}[V_{\Gamma_1 \cdot \Gamma_2}] \longrightarrow \mathrm{C}^{\mathrm{sing}}[V_{\Gamma_1}, V_{\Gamma_2}].$$

By the second part of Lemma 5.5.2, π and π' are oriented SA bundles and ρ induces a map of degree ± 1 between their fibers. It is easy to check that it is actually of degree $+1$ because it preserves their orientations (which depend, when N is odd, on the linear order of $I_{\Gamma_1} \obslash I_{\Gamma_2}$). Therefore by [**18**, Proposition 8.10], for any minimal form $\mu \in \Omega_{\min}(P)$, we have

$$\pi'_*(\mu) = \pi_*(\rho^*(\mu)). \tag{9.9}$$

We have then

$$\begin{array}{rcl} \widehat{\mathrm{I}}(\Gamma_1 \cdot \Gamma_2) & = & \pi_*(\theta^*_{\Gamma_1 \cdot \Gamma_2}(\mathrm{vol}_{E_{\Gamma_1} \obslash E_{\Gamma_2}})) \\ & = & \pi_*(\rho^*(q_1^*\theta^*_{\Gamma_1}(\mathrm{vol}_{E_{\Gamma_1}}) \wedge q_2^*\theta^*_{\Gamma_2}(\mathrm{vol}_{E_{\Gamma_2}}))) \\ & \overset{\text{Equation (9.9)}}{=} & \pi'_*\left(q_1^*\theta^*_{\Gamma_1}(\mathrm{vol}_{E_{\Gamma_1}}) \wedge q_2^*\theta^*_{\Gamma_2}(\mathrm{vol}_{E_{\Gamma_2}})\right) \\ & \overset{[\mathbf{18},\ \text{Proposition 8.15}]}{=} & \pi_{1*}(\theta^*_{\Gamma_1}(\mathrm{vol}_{E_{\Gamma_1}})) \wedge \pi_{2*}(\theta^*_{\Gamma_2}(\mathrm{vol}_{E_{\Gamma_2}})) \\ & = & \widehat{\mathrm{I}}(\Gamma_1) \cdot \widehat{\mathrm{I}}(\Gamma_2). \end{array}$$

□

9.3. Vanishing of $\widehat{\mathrm{I}}$ on non-admissible diagrams

Recall from Definition 6.5.1 the ideal $\mathcal{N}(A)$ of non-admissible diagrams. In this section we prove

PROPOSITION 9.3.1. $\widehat{\mathrm{I}}(\mathcal{N}(A)) = 0$.

REMARK 9.3.2. The ideal $\mathcal{N}(A)$ is not the entire kernel of $\widehat{\mathrm{I}}$ since for example there are admissible diagrams of arbitrarily high degrees but $\Omega^*_{PA}(\mathrm{C}[A])$ is bounded above.

Since $\mathcal{D}(A) = \widehat{\mathcal{D}}(A)/\mathcal{N}(A)$, we deduce the following

COROLLARY 9.3.3. $\widehat{\mathrm{I}}$ *induces a map of algebras*

$$\mathrm{I}\colon \mathcal{D}(A) \longrightarrow \Omega_{PA}(\mathrm{C}[A]).$$

DEFINITION 9.3.4. The maps

$$\widehat{\mathrm{I}}\colon \widehat{\mathcal{D}}(A) \longrightarrow \Omega_{PA}(\mathrm{C}[A]) \quad \text{and} \quad \mathrm{I}\colon \mathcal{D}(A) \longrightarrow \Omega_{PA}(\mathrm{C}[A]).$$

are called the *Kontsevich configuration space integrals.*

The proof of Proposition 9.3.1 consists of Lemmas 9.3.5–9.3.9.

LEMMA 9.3.5. $\widehat{\mathrm{I}}$ *vanishes on diagrams with loops.*

PROOF. If $|A| \leq 1$ the lemma is obvious. Suppose that $|A| \geq 2$ and let Γ be a diagram with a loop. One of the components of the map θ_Γ to the product $(S^{N-1})^{E_\Gamma}$ is a constant map. Therefore θ_Γ factors through a space of dimension $< (N-1)\cdot|E_\Gamma|$. By [**18**, Proposition 5.24] we deduce that the pullback of the maximal degree form vol_{E_Γ} by θ_Γ is zero, and hence the same is true for $\widehat{\mathrm{I}}(\Gamma)$. □

LEMMA 9.3.6. $\widehat{\mathrm{I}}$ *vanishes on diagrams with double edges.*

PROOF. If $|A| \leq 1$ the lemma is obvious. Suppose that $|A| \geq 2$ and let Γ be a diagram with double edges. The two components of the map θ_Γ corresponding to the double edges factor through the diagonal map

$$\Delta\colon S^{N-1} \longrightarrow S^{N-1} \times S^{N-1}.$$

Therefore θ_Γ factors through a space of dimension $< (N-1)\cdot|E_\Gamma|$. The conclusion is the same as in the proof of Lemma 9.3.5. □

LEMMA 9.3.7. $\widehat{\mathrm{I}}$ *vanishes on diagrams containing an internal vertex not connected to any external vertices.*

PROOF. The lemma is trivial if $|A| \leq 1$. Assume that $|A| \geq 2$. Let Γ be a diagram as in the statement. We have a factorization $\Gamma = \Gamma_1 \cdot \Gamma_2$ where Γ_1 is a diagram with at least one internal vertex and such that all edges are between internal vertices. Since $\widehat{\mathrm{I}}$ is a morphism of algebras, it is enough to prove that $\widehat{\mathrm{I}}(\Gamma_1) = 0$. So without loss of generality we assume that $\Gamma = \Gamma_1$.

The canonical projection π_Γ factors as

$$\mathrm{C}[V_\Gamma] \xrightarrow{\rho} \mathrm{C}[I_\Gamma] \times \mathrm{C}[A] \xrightarrow{q} \mathrm{C}[A]$$

where ρ is induced by the canonical projections on each factors, and q is the projection on the second factor. Since we have assumed that the edges of Γ are only between internal vertices, there is a factorization $\theta_\Gamma = \theta' \circ \rho$ for some map

$$\theta'\colon \mathrm{C}[I_\Gamma] \times \mathrm{C}[A] \longrightarrow (S^{N-1})^{E_\Gamma}.$$

Since Γ contains at least one internal vertex, Proposition 5.1.2 implies that

$$\dim(\mathrm{C}[I_\Gamma]) \leq N \cdot |I_\Gamma| - N.$$

Therefore for $x \in \mathrm{C}[A]$ we have

$$\dim(q^{-1}(x)) < N \cdot |I_\Gamma| = \dim(\pi_\Gamma^{-1}(x)))$$

and [**18**, Proposition 8.14] implies that

$$\widehat{\mathrm{I}}(\Gamma) = \pi_{\Gamma *}(\theta_\Gamma(\mathrm{vol}_{E_\Gamma})) = \pi_{\Gamma *}(\rho^*(\theta'^*(\mathrm{vol}_{E_\Gamma}))) = 0.$$

□

LEMMA 9.3.8. *$\widehat{\mathrm{I}}$ vanishes on diagrams containing a univalent internal vertex.*

PROOF. If $|A| \leq 1$, lemma is trivial. Suppose that $|A| \geq 2$. Let Γ be a diagram with an internal vertex i of valence 1 and let v be the only vertex adjacent to i. Then V_Γ has at least three vertices. Consider the projection

$$\rho\colon \mathrm{C}[V_\Gamma] \longrightarrow \mathrm{C}[\{i,v\}] \times \mathrm{C}[V_\Gamma \setminus \{i\}]$$

induced by the canonical projections on each factor. Since (i,v) is the only edge with endpoint i we have a factorization $\theta_\Gamma = \theta' \circ \rho$ for some map

$$\theta'\colon \mathrm{C}[\{i,v\}] \times \mathrm{C}[V_\Gamma \setminus \{i\}] \longrightarrow (S^{N-1})^{E_\Gamma}.$$

Since i is internal we get a map

$$q\colon \mathrm{C}[\{i,v\}] \times \mathrm{C}[V_\Gamma \setminus \{i\}] \longrightarrow \mathrm{C}[A]$$

obtained as the projection on the second factor followed by the canonical projection, and $\pi_\Gamma = q \circ \rho$. For $x \in \mathrm{C}[A]$,

$$\dim(q^{-1}(x)) < \dim(\pi_\Gamma^{-1}(x)).$$

Then [**18**, Proposition 8.14] implies that

$$\widehat{\mathrm{I}}(\Gamma) = \pi_{\Gamma *}(\theta_\Gamma(\mathrm{vol}_{E_\Gamma})) = \pi_{\Gamma *}(\rho^* \theta'^*(\mathrm{vol}_{E_\Gamma})) = 0.$$

□

LEMMA 9.3.9. *$\widehat{\mathrm{I}}$ vanishes on diagrams containing a bivalent internal vertex.*

PROOF. Lemma is trivial when $|A| \leq 1$. Assume that $|A| \geq 2$. We will use Kontsevich's trick from [**19**, Lemma 2.1]. Let Γ be a diagram with an internal vertex i of valence 2 and let v and w be its adjacent vertices. The key idea will be to consider the automorphism of $\mathrm{C}[V_\Gamma]$ which replaces the point labeled by i by a point symmetric to it with respect to the barycenter of the points labeled by v and w, and to use this symmetry to show that $\widehat{\mathrm{I}}(\Gamma)$ is equal to its negative. For concreteness, suppose that the two edges at i are oriented as (v,i) and (w,i), and ordered by $(v,i) < (w,i)$ as the last two edges of the ordered set E_Γ.

To give the idea of the proof suppose first that the diagram consists only of these three vertices and two edges, with v and w external. Set $\theta = (\theta_{v,i}, \theta_{w,i})$, which in this special case is exactly θ_Γ, and set $\pi = \pi_\Gamma$.

Consider the continuous involution

$$\chi\colon \mathrm{C}[\{v,w,i\}] \xrightarrow{\cong} \mathrm{C}[\{v,w,i\}]$$

defined on $\mathrm{C}(\{v,w,i\})$ by

$$\chi(y) = (y(v)\,,\, y(w)\,,\, y(v) + y(w) - y(i))$$

where $y(v)+y(w)-y(i)$ is the point symmetric to $y(i)$ with respect to the barycenter $y(v)$ and $y(w)$. This is a semi-algebraic automorphism of degree $(-1)^N$.

Let

$$A\colon S^{N-1} \longrightarrow S^{N-1}$$

be the antipodal map and let

$$\tau\colon S^{N-1}\times S^{N-1} \longrightarrow S^{N-1}\times S^{N-1} \tag{9.10}$$

be the interchange of factors which is of degree $(-1)^{N-1}$. By construction of χ, the following diagram commutes

$$\begin{array}{ccc} C[\{v,w,i\}] & \xrightarrow{\theta} & S^{N-1}\times S^{N-1} \\ \chi\downarrow & & \downarrow \tau\circ(A\times A) \\ C[\{v,w,i\}] & \xrightarrow{\theta} & S^{N-1}\times S^{N-1} \end{array} \tag{9.11}$$

By symmetry of vol, we have $A^*(\mathrm{vol}) = \pm\,\mathrm{vol}$, so

$$(\tau\circ(A\times A))^*(\mathrm{vol}\times\mathrm{vol}) = (-1)^{N-1}(\mathrm{vol}\times\mathrm{vol})$$

and hence

$$\chi^*\theta^*(\mathrm{vol}_{E_\Gamma}) = (-1)^{N-1}\theta^*(\mathrm{vol}_{E_\Gamma}). \tag{9.12}$$

On the other hand the restriction of χ to each fiber $\pi^{-1}(x)$, $x\in \mathrm{C}[A]$, is an SA homeomorphism of degree $(-1)^N$. By [**18**, Proposition 8.10],

$$\pi_*(\chi^*(\theta^*(\mathrm{vol}_{E_\Gamma}))) = (-1)^N\pi_*(\theta^*(\mathrm{vol}_{E_\Gamma})). \tag{9.13}$$

We deduce that

$$\begin{array}{rcl} \widehat{\mathrm{I}}(\Gamma) & = & \pi_*(\theta^*\,\mathrm{vol}_{E_\Gamma}) \\ & \overset{\text{Equation (9.13)}}{=} & (-1)^N\pi_*(\chi^*\theta^*\,\mathrm{vol}_{E_\Gamma}) \\ & \overset{\text{Equation (9.12)}}{=} & (-1)^{N-1}(-1)^N\pi_*(\theta^*\,\mathrm{vol}_{E_\Gamma}) \\ & = & -\widehat{\mathrm{I}}(\Gamma), \end{array}$$

and hence $\widehat{\mathrm{I}}(\Gamma)=0$.

For the case of a general diagram, consider the fiber product

$$\begin{array}{ccc} P & \longrightarrow & \mathrm{C}[\{v,w,i\}] \\ \downarrow & \text{pullback} & \downarrow \pi_1 \\ \mathrm{C}[V_\Gamma\setminus\{i\}] & \xrightarrow[\pi_2]{} & \mathrm{C}[\{v,w\}] \end{array} \tag{9.14}$$

where π_1 and π_2 are the canonical projections. Since $\pi_1\circ\chi=\pi_1$, the automorphism χ of $\mathrm{C}[\{v,w,i\}]$ can be mixed with the identity map on on $\mathrm{C}[V_\Gamma\setminus\{i\}]$ to give an automorphism of P that we also denote by χ. The canonical projections

$$\mathrm{C}[V_\Gamma]\longrightarrow \mathrm{C}[V_\Gamma\setminus\{i\}] \quad\text{and}\quad \mathrm{C}[V_\Gamma]\longrightarrow \mathrm{C}[\{v,w,i\}]$$

induce a map $\rho\colon \mathrm{C}[V_\Gamma]\to P$. We have a factorization $\pi_\Gamma = \pi\circ\rho$ for some map $\pi\colon P\to \mathrm{C}[A]$ which is an oriented SA bundle.

Since the only edges with endpoint i are (v,i) and (w,i), there is a factorization $\theta_\Gamma = \theta \circ \rho$ for some map

$$\theta\colon P \longrightarrow \left(S^{N-1}\right)^{E_\Gamma \setminus \{(v,i),(w,i)\}} \times S^{N-1} \times S^{N-1}.$$

For each $x \in \mathrm{C}[A]$ the restriction of ρ to the interior of $\pi_\Gamma^{-1}(x)$ is an oriented homeomorphism onto a dense image in the fiber $\pi^{-1}(x)$. By naturality of integration along the fiber [**18**, Proposition 8.10],

$$\pi_{\Gamma *}(\theta_\Gamma^*(\mathrm{vol}_{E_\Gamma})) = \pi_*(\theta^*(\mathrm{vol}_{E_\Gamma})). \tag{9.15}$$

As for the Diagram (9.11), we have $\theta \circ \chi = (\mathrm{id} \times \tau \circ (A \times A)) \circ \theta$. The rest of the proof is the same as in the special case treated above, starting with Equation (9.12). □

PROOF OF PROPOSITION 9.3.1. A non admissible diagram satisfies the hypothesis of one of Lemmas 9.3.5–9.3.9. □

9.4. $\hat{\mathrm{I}}$ and I are chain maps

This section is devoted to the proof of the following.

PROPOSITION 9.4.1. *The Kontsevich configuration space integrals commute with the differential, that is,*

$$\widehat{\mathrm{I}}\, d = d\, \widehat{\mathrm{I}} \quad \textit{and} \quad \mathrm{I}\, d = d\, \mathrm{I}.$$

Let A be a finite set and let Γ be a diagram on A. We will prove that $\widehat{\mathrm{I}}(d(\Gamma)) = d(\widehat{\mathrm{I}}(\Gamma))$, which by Corollary 9.3.3 implies the analogous result for I. If $|A| \le 1$ then this is obvious. Also if Γ is non-admissible, then by Proposition 9.3.1 and Lemma 6.5.2 we have

$$d(\widehat{\mathrm{I}}(\Gamma)) = 0 = \widehat{\mathrm{I}}(d(\Gamma)).$$

So now we assume that $|A| \ge 2$ and that Γ is admissible.

From now on we will drop Γ from the notation when it appears as an index, so $\Gamma = (A, I, E, s, t)$, $V := V_\Gamma$, $\pi := \pi_\Gamma$, etc. Also, to easily define orientations of certain configuration spaces we assume that A is equipped with an arbitrary linear order and that $V = A \otimes I$.

On one side, by definition of $d(\Gamma)$ in (6.1),

$$\widehat{\mathrm{I}}(d(\Gamma)) = \sum_{e \in E^{\mathrm{contr}}} \epsilon(\Gamma, e) \cdot \widehat{\mathrm{I}}(\Gamma/e). \tag{9.16}$$

To develop the other side $d(\widehat{\mathrm{I}}(\Gamma))$, we will need the results from Sections 5.4 and 5.7 on the decomposition of the fiberwise boundary of $\mathrm{C}[V]$ into faces which are the images of operadic maps Φ_W defined in (5.11), mainly Propositions 5.4.1 and 5.7.1. Recall from (5.37) the fiberwise boundary of π,

$$\pi^\partial\colon \mathrm{C}^\partial[V] \longrightarrow \mathrm{C}[A].$$

Since $\widehat{\mathrm{I}}(\Gamma) = \pi_*(\theta^*(\mathrm{vol}_E))$ and $\theta^*(\mathrm{vol}_E)$ is a cocycle, the fiberwise Stokes formula of [**18**, Proposition 8.12] implies that

$$d(\widehat{\mathrm{I}}(\Gamma)) = (-1)^{\deg(\Gamma)} \cdot \pi_*^\partial\left((\theta^* \mathrm{vol}_E)|\, \mathrm{C}^\partial[V]\right), \tag{9.17}$$

where $(\theta^* \mathrm{vol}_E)|\, \mathrm{C}^\partial[V]$ denotes the restriction of the form $\theta^* \mathrm{vol}_E$ to that subspace. Set

$$\mu := (\theta^* \mathrm{vol}_E)|\, \mathrm{C}^\partial[V] \in \Omega^*_{\min}(\mathrm{C}^\partial[V]).$$

Using the decomposition of the fiberwise boundary of $\mathrm{C}[V]$ from Proposition 5.7.1 and Proposition 5.4.1 (ii)-(iii), we get, by additivity of integration along the fiber [**18**, Proposition 8.11],

$$\pi^{\partial}_{*}(\mu) = \sum_{W \in \mathcal{BF}(V,A)} (\pi^{\partial}|\operatorname{im}\Phi_W)_*(\mu) \tag{9.18}$$

with the notation from Sections 5.4 and 5.7. Recall in particular that $\mathcal{BF}(V, A)$ is the indexing set of some faces of the fiberwise boundary and consist of some subsets $W \subset V$.

The core of the proof of Proposition 9.4.1 consists of computing the terms of the sum in (9.18). We will prove that they all vanish except when W is the pair of endpoints of a contractible edge e of Γ, and in that case

$$(\pi^{\partial}|\operatorname{im}\Phi_W)_*(\mu) = \pm\widehat{\mathrm{I}}(\Gamma/e),$$

which are exactly the terms of $\widehat{\mathrm{I}}(d(\Gamma))$ in (9.16).

Let $W \in \mathcal{BF}(V, A)$, that is: $W \subsetneq V$, $|W| \geq 2$, and either $A \subset W$ or $|W \cap A| \leq 1$ (see (5.38)). Consider the projection to the quotient set

$$q \colon V \longrightarrow V/W.$$

The composite

$$(V \setminus W) \cup \{\min(W)\} \hookrightarrow V \xrightarrow{q} V/W \tag{9.19}$$

is a bijection and we use it to transport the linear order of V to V/W.

In order to compute $(\pi^{\partial}|\operatorname{im}\Phi_W)_*(\mu)$ in (9.21) below, we first associate to Γ and W two diagrams: Γ' which is the full subgraph of Γ with set of vertices W, and $\overline{\Gamma}$ which is the quotient of Γ by the subgraph Γ'. More precisely, $\Gamma' := (A', I', E', s', t')$ where

- $A' := A \cap W$;
- $I' := I \cap W$;
- $E' := E \cap s^{-1}(W) \cap t^{-1}(W)$;
- $s' = s|E'$ and $t' = t|E'$,

and $\overline{\Gamma} := (\overline{A}, \overline{I}, \overline{E}, \overline{s}, \overline{t})$ with

- $\overline{A} := q(A)$
- $\overline{I} := (V/W) \setminus q(A)$;
- $\overline{E} := E \setminus E'$;
- $\overline{s} = q \circ (s|\overline{E})$ and $\overline{t} = q \circ (t|\overline{E})$.

Hence $V_{\Gamma'} = W$ and $V_{\overline{\Gamma}} = V/W$.

Set $\overline{\theta} := \theta_{\overline{\Gamma}}$ and $\theta' := \theta_{\Gamma'}$. Set also the minimal forms $\overline{\mu} = \overline{\theta}^*(\mathrm{vol}_{\overline{E}})$ and $\mu' = \theta'^*(\mathrm{vol}_{E'})$. The following diagram is commutative

$$\begin{array}{ccccc}
(S^{N-1})^{\overline{E}} \times (S^{N-1})^{E'} & & \xrightarrow{\ \tau_W\ \cong\ } & & (S^{N-1})^{E} \\
\uparrow{\scriptstyle \overline{\theta}\times\theta'} & & & & \uparrow{\scriptstyle \theta} \\
\mathrm{C}[V/W] \times \mathrm{C}[W] & \xrightarrow{\ \Phi_W\ } & \mathrm{C}^{\partial}[V] & \hookrightarrow & \mathrm{C}[V] \\
& {\scriptstyle \pi^{\partial}\circ\Phi_W}\searrow & \downarrow{\scriptstyle \pi^{\partial}} & \swarrow{\scriptstyle \pi} & \\
& & \mathrm{C}[A] & &
\end{array} \tag{9.20}$$

Here τ_W is the obvious reordering of factors which is a homeomorphism since $E = \overline{E} \amalg E'$.

Since $W \in \mathcal{BF}(V, A)$, there are two cases:

(1) $A \subset W$. Then we have a canonical projection
$$\pi' \colon \mathrm{C}[W] \longrightarrow \mathrm{C}[A],$$
and $\pi^\partial \circ \Phi_W = \pi' \circ \mathrm{proj}_2$ where $\mathrm{proj}_2 \colon \mathrm{C}[V/W] \times \mathrm{C}[W] \to \mathrm{C}[W]$ is the projection on the second factor.

(2) $|W \cap A| \leq 1$. Then the composite
$$A \hookrightarrow V \xrightarrow{q} V/W$$
is injective, and we have an associated canonical projection
$$\overline{\pi} \colon \mathrm{C}[V/W] \longrightarrow \mathrm{C}[A].$$
Further, $\pi^\partial \circ \Phi_W = \overline{\pi} \circ \mathrm{proj}_1$ where proj_1 is the projection on the first factor.

In both cases, $\pi^\partial \circ \Phi_W$ is the composition of two oriented SA bundles, and hence is itself an oriented SA bundle [**18**, Proposition 8.5].

The linear orders on V/W and W give $\mathrm{C}[V/W] \times \mathrm{C}[W]$ a natural orientation, as well as to the fibers of $\pi^\partial \circ \Phi_W$. Define the sign
$$\mathrm{sign}(\Phi_W) = \pm 1$$
according to whether
$$\Phi_W \colon \mathrm{C}[V/W] \times \mathrm{C}[W] \longrightarrow \mathrm{C}^\partial[V],$$
which is a homeomorphism onto its image of codimension 0, preserves or reverses orientation. Then Φ_W induces the same change of orientation between the fibers over any $x \in \mathrm{C}[A]$. Define also $\mathrm{sign}(\tau_W) = \pm 1$ by
$$\tau_W^*(\mathrm{vol}_E) = \mathrm{sign}(\tau_W) \cdot (\mathrm{vol}_{\overline{E}} \times \mathrm{vol}_{E'}).$$

The Diagram (9.20) and [**18**, Proposition 8.10] imply that

$$(\pi^\partial|\,\mathrm{im}\,\Phi_W)_*(\mu) = \mathrm{sign}(\Phi_W) \cdot \mathrm{sign}(\tau_W) \cdot \left((\pi^\partial \circ \Phi_W)_*(\overline{\mu} \times \mu')\right). \tag{9.21}$$

Our computation of $(\pi^\partial|\,\mathrm{im}\,\Phi_W)_*(\mu)$ goes through the following lemma, in which we use the notation $\langle \omega, [\![M]\!] \rangle$ to denote the evaluation on a compact oriented semi-algebraic manifold M of a PA form $\omega \in \Omega_{PA}(M)$ (see equations (4.1) and (4.4)); in other words
$$\langle \omega, [\![M]\!] \rangle = \int_M \omega.$$

LEMMA 9.4.2.

$$(\pi^\partial \circ \Phi_W)_*(\overline{\mu} \times \mu') = \begin{cases} \overline{\pi}_*(\overline{\mu}) \cdot \langle \mu', [\![\mathrm{C}[W]]\!] \rangle, & \text{if } |W \cap A| \leq 1; \\ \pm \pi'_*(\mu') \cdot \langle \overline{\mu}, [\![\mathrm{C}[V/W]]\!] \rangle, & \text{if } A \subset W. \end{cases} \tag{9.22}$$

PROOF. If $|W \cap A| \leq 1$ then $\pi^\partial \circ \Phi_W = \overline{\pi} \circ \mathrm{proj}_1$ and the desired formula is a consequence of the double pushforward formula of [**18**, Proposition 8.13].

If $A \subset W$ then $\pi^\partial \circ \Phi_W = \pi' \circ \mathrm{proj}_2$ and the desired formula is again a consequence of the double pushforward formula, with an extra sign because of the interchange of factors in $\mathrm{C}[V/W] \times \mathrm{C}[V]$ to apply the double pushforward formula. □

Our next task is to show that in the right hand side of (9.22), the expressions

$$\langle \overline{\mu} \,,\, [\![\mathrm{C}[V/W]]\!] \rangle \quad \text{and} \quad \langle \mu' \,,\, [\![\mathrm{C}[W]]\!] \rangle$$

vanish, except when W is the pair of endpoints of a contractible edge. This is the content of Lemmas 9.4.5–9.4.7. To prove them we first establish the following general vanishing lemma.

LEMMA 9.4.3. *Let Γ_0 be a diagram with at least 3 vertices. Then*

$$\langle \theta^*_{\Gamma_0}(\mathrm{vol}_{E_{\Gamma_0}}) \,,\, [\![\mathrm{C}[V_{\Gamma_0}]]\!] \rangle \;=\; 0. \tag{9.23}$$

PROOF. In this proof we drop Γ_0 from the notation when it appears as an index, so here $V := V_{\Gamma_0}$, $E := E_{\Gamma_0}$, and $\theta := \theta_{\Gamma_0}$. By hypothesis, $|V| \geq 3$.

We can assume that

$$\deg \theta^*(\mathrm{vol}_E) = \dim \mathrm{C}[V] \tag{9.24}$$

because otherwise the left side of (9.23) vanishes for degree reasons.

If Γ_0 has an isolated vertex v then θ factors through $\mathrm{C}[V \setminus \{v\}]$. Since

$$\dim \mathrm{C}[V \setminus \{v\}] < \dim \mathrm{C}[V],$$

the left side of (9.23) again vanishes for degree reasons.

If Γ_0 has a univalent vertex and $|V| \geq 3$ then the left side of (9.23) vanishes by the same argument as in the main part of the proof of Lemma 9.3.8 (where the relevant hypothesis is that there are at least three vertices).

If Γ_0 has a bivalent vertex then the vanishing follows by the same argument as in Lemma 9.3.9.

Finally, suppose that all the vertices of Γ_0 are at least trivalent.

If $N = 2$ then $|E| \geq 3$ and the statement is exactly that of [**21**, Lemma 6.4].

Suppose that $N \geq 3$. Since all the vertices are at least trivalent, $|E| \geq \frac{3}{2}|V|$. Therefore

$$\begin{aligned}
\deg(\theta^*(\mathrm{vol}_E)) = (N-1) \cdot |E| &\geq \frac{3(N-1)}{2}|V| \\
&= N \cdot |V| + \frac{N-3}{2} \cdot |V| \\
&\geq N \cdot |V| > \dim \mathrm{C}[V]
\end{aligned}$$

which contradicts Equation (9.24). □

REMARK 9.4.4. The above proof is essentially the one given in [**8**, Appendix A.3]. However, the context is different in that situation since the configuration space integrals produce differential forms on the spaces of knots rather then on configuration spaces, as is the case here.

LEMMA 9.4.5. *If $A \subset W$, then $\langle \overline{\mu} \,,\, [\![\mathrm{C}[V/W]]\!] \rangle = 0$.*

PROOF. If $|V/W| \geq 3$ then we apply Lemma 9.4.3 to $\Gamma_0 = \overline{\Gamma}$.

Otherwise $|V/W| = 2$ and $V = W \cup \{v\}$ for some internal vertex v of Γ. Since Γ is admissible, v is at least trivalent and its adjacent vertices are in W. Therefore $\overline{\Gamma}$ has double edges (even triple) and the conclusion is the same as in the proof of Lemma 9.3.6. □

LEMMA 9.4.6. *If $|W| \geq 3$ or if W is a pair of non-adjacent vertices of Γ, then*

$$\langle \mu' \,,\, [\![\mathrm{C}[W]]\!] \rangle \;=\; 0.$$

PROOF. If $|W| \geq 3$, apply Lemma 9.4.3 to $\Gamma_0 = \Gamma'$.

If W is a pair of non adjacent vertices, then Γ' has no edges and hence $\mu' = 1 \in \Omega^0_{\min}(\mathrm{C}[W])$. As $N > 1$, $\deg(\mu') = 0 < \dim \mathrm{C}[W]$ and the statement follows. □

We are finally left with the case when W is a pair of adjacent vertices of Γ and $|W \cap A| \leq 1$. Then the edge e connecting these two vertices is contractible because at most one of the endpoints is external and it is not a loop nor a dead end since Γ is admissible. Moreover in that case we have

$$\overline{\Gamma} = \Gamma/e \quad \text{and} \quad \overline{\pi}_*(\overline{\mu}) = \widehat{\mathrm{I}}(\Gamma/e).$$

(The order of internal vertices in $\overline{\Gamma}$ is the same as for Γ/e because the ordering (9.19) is compatible with that of $I_{\Gamma/e}$ from Definition 6.4.1.) Define the sign

$$\eta(e) = \begin{cases} +1, & \text{if } N \text{ is even or } s(e) < t(e) \\ -1, & \text{otherwise.} \end{cases} \tag{9.25}$$

LEMMA 9.4.7. *If W is a pair of vertices connected by a contractible edge e of Γ then*

$$\langle \mu', [\![\mathrm{C}[W]]\!] \rangle = \eta(e).$$

PROOF. Γ' consists of a single edge and we have a homeomorphism

$$\theta' = \theta_{s(e),t(e)} \colon \mathrm{C}[\{s(e), t(e)\}] \longrightarrow S^{N-1}$$

which preserves or reverses orientation according to the sign $\eta(e)$. Thus

$$\langle \mu', [\![\mathrm{C}[W]]\!] \rangle = \eta(e) \cdot \int_{S^{N-1}} \mathrm{vol} = \eta(e).$$

□

Also set $\Phi_e = \Phi_W$ and $\tau_e = \tau_W$ in that case.

Collecting (9.17), (9.18), (9.21), Lemma 9.4.2, and Lemmas 9.4.5–9.4.7, we get

$$d(\widehat{\mathrm{I}}(\Gamma)) = \sum_{e \in E^{\mathrm{contr}}} (-1)^{\deg(\Gamma)} \cdot \mathrm{sign}(\Phi_e) \cdot \mathrm{sign}(\tau_e) \cdot \eta(e) \cdot \widehat{\mathrm{I}}(\Gamma/e). \tag{9.26}$$

Comparing this to the formula (9.16) for $\widehat{\mathrm{I}}(d(\Gamma))$, it remains to compare the signs of the terms in (9.26) and (9.16). Let e be a contractible edge of Γ.

LEMMA 9.4.8. $\mathrm{sign}(\tau_e) = (-1)^{(N-1)\cdot(\mathrm{pos}(e:E)+|E|)}$.

PROOF. If e is the last edge in the order of E then τ_e is the identity map, and hence $\mathrm{sign}(\tau_e) = +1$ which is the expected value since $\mathrm{pos}(e : E) = |E|$.

When one transposes e with a consecutive edge in the linear order of E then both $\mathrm{sign}(\tau_e)$ and $(-1)^{(N-1)\cdot(\mathrm{pos}(e:E)+|E|)}$ change by a factor of $(-1)^{N-1}$. This proves the lemma in full generality. □

LEMMA 9.4.9. $\mathrm{sign}(\Phi_e) = (-1)^{N\cdot(\mathrm{pos}(\max(s(e),t(e)):I)+|I|)}$.

PROOF. Suppose first that $t(e)$ is the last and $s(e)$ the second to the last vertex in the linear order of $A \otimes I$. Then it is easy to see that

$$\Phi_e \colon \mathrm{C}[V \setminus \{t(e)\}] \times \mathrm{C}[\{s(e), t(e)\}] \longrightarrow \partial\, \mathrm{C}[V]$$

is orientation-preserving, and hence

$$\mathrm{sign}(\Phi_e) = +1 = (-1)^{N\cdot(|I|+|I|)}$$

as expected.

Consider now a permutation σ of the set of vertices and its induced action on the following diagram

$$\begin{array}{ccc} \mathrm{C}[V\setminus\{\max(s(e),t(e))\}]\times \mathrm{C}[\{s(e),t(e)\}] & \xrightarrow{\Phi_{\{s(e),t(e)\}}} & \partial\,\mathrm{C}[V] \\ \downarrow{\scriptstyle \sigma\times\sigma} & & \downarrow{\scriptstyle \sigma} \\ \mathrm{C}[V\setminus\{\max(\sigma(s(e)),\sigma(t(e)))\}]\times \mathrm{C}[\{\sigma(s(e)),\sigma(t(e))\}] & \xrightarrow{\Phi_{\sigma(s(e)),\sigma(t(e))\}}} & \partial\,\mathrm{C}[V]. \end{array}$$

Inspecting the changes of signs through this diagram, it is straighforward to check that the formula is true in general. □

By Lemmas 9.4.8–9.4.9 we get that the expressions at (9.26) and (9.16) are equal. This finishes the proof of Proposition 9.4.1 showing that $\widehat{\mathrm{I}}$ and I are chain maps.

9.5. $\hat{\mathrm{I}}$ and I are almost morphisms of cooperads

Ideally, $\widehat{\mathrm{I}}$ and I would be morphisms of cooperads. However, as explained in Chapter 3, this is not true since $\Omega_{PA}(\mathrm{C}[\bullet])$ is not a cooperad because Ω_{PA} is not comonoidal. However, these maps are almost morphisms of cooperads in the following sense.

PROPOSITION 9.5.1. *The Kontsevich configuration space integrals $\widehat{\mathrm{I}}$ and I are compatible with the cooperad structures on $\widehat{\mathcal{D}}$ and $\mathcal{D}$ as well as with the structure induced on $\Omega_{PA}(\mathrm{C}[\bullet])$ by the operad structure on $\mathrm{C}[\bullet]$. Namely, we have*

(1) *Given a weak ordered partition $\nu\colon A\to P$, set $P^*=\{0\}\otimes P$, $A_p=\nu^{-1}(p)$, and $A_0=P$ as in the setting 2.4.1. Recall the (co)operad structure maps*

$$\Phi_\nu\colon \prod_{p\in P^*}\mathrm{C}[A_p]\longrightarrow \mathrm{C}[A]\,,$$
$$\widehat{\Psi}_\nu\colon \widehat{\mathcal{D}}(A)\longrightarrow \underset{p\in P^*}{\otimes}\widehat{\mathcal{D}}(A_p).$$

Then the following diagram is commutative:

$$\begin{array}{ccc} \widehat{\mathcal{D}}(A) & \xrightarrow{\ \widehat{\mathrm{I}}\ } & \Omega_{PA}(\mathrm{C}[A]) \\ & & \downarrow{\scriptstyle \Phi_\nu^*=\Omega_{PA}(\Phi_\nu)} \\ \downarrow{\scriptstyle \widehat{\Psi}_\nu} & & \Omega_{PA}(\prod_{p\in P^*}\mathrm{C}[A_p]) \\ & & \uparrow{\scriptstyle \simeq\ \times} \\ \otimes_{p\in P^*}\widehat{\mathcal{D}}(A_p) & \xrightarrow[\otimes_{p\in P^*}\widehat{\mathrm{I}}]{} & \otimes_{p\in P^*}\Omega_{PA}(\mathrm{C}[A_p]) \end{array}$$

Here the right vertical quasi-isomorphism $\times$ is the standard Kunneth quasi-isomorphism on forms;

(2) *$\widehat{\mathrm{I}}$ is equivariant with respect to the action of the permutations of A;*

(3) *$\widehat{\mathrm{I}}$ commutes with the counits $\hat{\eta}\colon\widehat{\mathcal{D}}(1)\to\mathbb{R}$ and $\Omega_{PA}(C[1])\xrightarrow{\cong}\mathbb{R}$.*

(1)–(3) are also true when we replace $\widehat{\mathrm{I}}$ by I, $\widehat{\mathcal{D}}$ by $\mathcal{D}$, $\widehat{\Psi}_\nu$ by Ψ_ν, and $\hat{\eta}$ by η.

The rest of the section is devoted to the proof of this proposition. Statements (2) and (3) are easy, as is (1) when $|A| \leq 1$, and that the statements pertaining to I follow from those pertaining to $\widehat{\mathrm{I}}$.

We now focus on the proof of (1) for a given weak ordered partition $\nu\colon A \to P$ such that $|A| \geq 2$. Let Γ be a diagram on A. We need to prove that

$$(\times_{p\in P^*}\widehat{\mathrm{I}})(\widehat{\Psi}_\nu(\Gamma)) = \Phi_\nu^*(\widehat{\mathrm{I}}(\Gamma)). \tag{9.27}$$

To understand why this formula holds, remember the discussion of condensations of configurations starting soon after (5.16) and ending at Definition 5.6.1. Morally, the right hand side of the formula is the restriction of the form $\widehat{\mathrm{I}}(\Gamma)$ to the part of the boundary of $\mathrm{C}[A]$ consisting of ν-condensed configurations (assuming that ν is non-degenerate.) When performing integration along the fiber of π_Γ over a ν-condensed configuration $x \in \mathrm{C}[A]$, the points of the configuration $y \in \pi_\Gamma^{-1}(x) \subset \mathrm{C}[V]$ labeled by internal vertices can be differently condensed with respect to the various clusters of points in x, and this corresponds exactly to the different condensations λ relative to ν. Thus the integral $\Phi_\nu^*(\widehat{\mathrm{I}}(\Gamma))$ is obtained by summing over various subdomains $\mathrm{C}[V,\lambda] \subset \mathrm{C}[V]$ indexed by condensations, and the cooperad structure map $\widehat{\Psi}_\nu$ that appears on the left side of (9.27) is precisely the sum over these condensations.

We now proceed with the details. To simplify notation, we will drop Γ from the notation when it appears as an index, so $I := I_\Gamma$, $\pi := \pi_\Gamma$, $E := E_\Gamma$, $\theta = \theta_\Gamma$, etc. Also for a given condensation λ of V relative to ν and for $p \in P^*$ we will replace the index $\Gamma(\lambda,p)$ by p, as in $V_p := V_{\Gamma(\lambda,p)}$, $\theta_p := \theta_{\Gamma(\lambda,p)}$, etc.

The proof of Equation (9.27) relies on the decomposition of the pullback of the operad structure map Φ_ν along the canonical projection π that we have investigated in Section 5.6. We will use the notation and results from that section. Thus consider the pullback $\mathrm{C}[V,\nu]$ of Φ_ν along π from Diagram (5.15). Recall from Proposition 5.6.2 that we have a decomposition

$$\mathrm{C}[V,\nu] = \cup\, \mathrm{C}[V,\lambda]$$

where λ runs over all (essential) condensations λ of ν, and that this decomposition is "almost" a partition (Proposition 5.6.6). Fix such a condensation λ and consider the following diagram, where the bottom left triangle is Diagram (5.31), the right bottom pullback is Diagram (5.15), and τ_λ is the obvious interchange of factors:

(9.28)

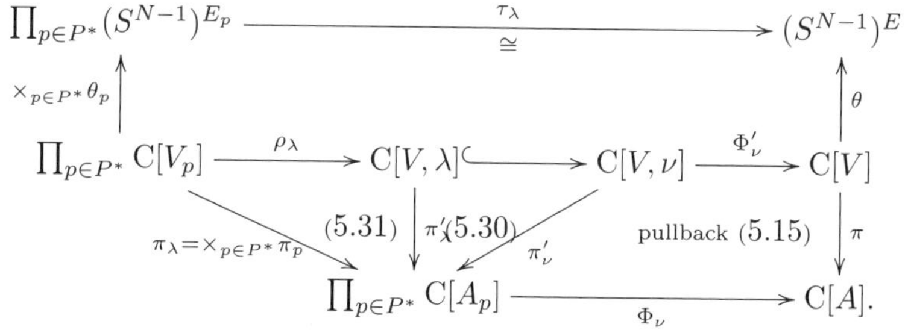
$\prod_{p\in P^*}(S^{N-1})^{E_p}$
τ_λ
$\cong$
$(S^{N-1})^E$
$\times_{p\in P^*}\theta_p$
θ
$\prod_{p\in P^*}\mathrm{C}[V_p]$
ρ_λ
$\mathrm{C}[V,\lambda]$
$\mathrm{C}[V,\nu]$
Φ'_ν
$\mathrm{C}[V]$
(5.31)
π'_λ(5.30)
$\pi_\lambda = \times_{p\in P^*}\pi_p$
π'_ν
pullback (5.15)
π
$\prod_{p\in P^*}\mathrm{C}[A_p]$
Φ_ν
$\mathrm{C}[A]$.

The top rectangle in this diagram is also commutative. Indeed by Proposition 5.6.5 (iii), $\Phi'_\nu \circ \rho_\lambda = \Phi'_\lambda$ and this can be identified with an operadic map (see (5.24) and (5.23)). From this it follows easily that the rectangle commutes. It is exactly in the commutativity of that rectangle that the compatibility between Φ_ν and $\widehat{\Psi}_\nu$ appears. Recall also from Proposition 5.6.5 that π_λ and π'_λ are oriented SA bundles and that ρ_λ induces a map of degree $\sigma(I,\lambda) = \pm 1$ between the fibers.

The idea of the proof of Equation (9.27) is to use this diagram to relate the left side of (9.27), which is the sum over all condensations λ of

$$\times_{p\in P^*}\widehat{\mathrm{I}}(\Gamma(\lambda,p)) = \times_{p\in P^*}\pi_{p*}(\theta_p^*\,\mathrm{vol}_{E_p}),$$

to the right side of (9.27), which is the pullback through Φ_ν^* of

$$\widehat{\mathrm{I}}(\Gamma) = \pi_*(\theta^*(\mathrm{vol}_E)).$$

To make this precise, recall the sign

$$\sigma(E,\lambda) = \pm 1$$

defined in (7.3) just before Lemma 7.1.3.

LEMMA 9.5.2. $\tau_\lambda^*(\mathrm{vol}_E) = \sigma(E,\lambda)\cdot\left(\times_{p\in P^*}\ \mathrm{vol}_{E_p}\right)$.

PROOF. Switching two factors of S^{N-1} is a map of degree $(-1)^{N-1}$. The factors of $\prod_{p\in P^*}(S^{N-1})^{E_p}$ are ordered as $\otimes_{p\in P^*}E_p$ and the number of transpositions needed to reorder this set as E is the cardinality of $S(E,\lambda)$. □

Recall from Definition 5.6.1 the notion of an essential condensation, which is a condensation λ such that, for each $p\in P^*$, $I_p=\emptyset$ (that is, $I_\Gamma\cap\lambda^{-1}(p)=\emptyset$) when $|A_p|\le 2$, and let

$$\mathrm{EssCond}(\Gamma) = \mathrm{EssCond}(V_\Gamma,\nu)$$

be the set of essential condensations of the diagram Γ.

LEMMA 9.5.3. *Let λ be a condensation of Γ.*

(i) If λ is essential, then for each $p\in P^*$

$$\widehat{\mathrm{I}}(\Gamma(\lambda,p)) = \pi_{p*}(\theta_p^*(\mathrm{vol}_{E_p})).$$

(ii) If λ is not essential, then

$$\left(\times_{p\in P^*}\widehat{\mathrm{I}}\right)(\Gamma(\lambda)) = 0.$$

PROOF. Suppose that λ is essential. Then for each $p\in P^*$, either $|A_p|\ge 2$, in which case $\widehat{\mathrm{I}}(\Gamma(\lambda,p))$ is given by the pushforward (9.6) as expected, or $I_p=\emptyset$ in which case formulas (9.7) and (9.6) agree because π_p is the identity map and $\mathrm{C}[A_p]=*$.

If λ is not essential then for some $p\in P^*$ we have $|A_p|\le 1$ and $I_p\ne\emptyset$, in which case $\widehat{\mathrm{I}}(\Gamma(\lambda,p))=0$ by (9.7). □

We can now prove the commutativity of the diagram in Proposition 9.5.1, part (1), which amounts to showing Equation (9.27). By inspection of Diagram (9.28) we have the following sequence of equalities. The justification for each equality is given at the end of the string.

$$\begin{aligned}
\Phi_\nu^*(\widehat{\mathrm{I}}(\Gamma)) &\overset{\text{(i)}}{=} \Phi_\nu^*(\pi_*(\theta^*(\mathrm{vol}_E))) \\
&\overset{\text{(ii)}}{=} \pi'_{\nu *}(\Phi_\nu'^*(\theta^* \,\mathrm{vol}_E)) \\
&\overset{\text{(iii)}}{=} \sum_{\lambda\in\mathrm{EssCond}(\Gamma)} \pi'_{\lambda *}((\Phi'_\nu|\, \mathrm{C}[V,\lambda])^*\theta^*(\mathrm{vol}_E)) \\
&\overset{\text{(iv)}}{=} \sum_{\lambda\in\mathrm{EssCond}(\Gamma)} \sigma(I,\lambda)\cdot \pi_{\lambda *}(\rho_\lambda^*(\Phi'_\nu|\, \mathrm{C}[V,\lambda])^*\theta^*(\mathrm{vol}_E)) \\
&\overset{\text{(v)}}{=} \sum_{\lambda\in\mathrm{EssCond}(\Gamma)} \sigma(I,\lambda)\cdot \pi_{\lambda *}\left((\times_{p\in P^*}\theta_p)^*(\tau_\lambda^*(\mathrm{vol}_E)\right) \\
&\overset{\text{(vi)}}{=} \sum_{\lambda\in\mathrm{EssCond}(\Gamma)} \sigma(I,\lambda)\cdot\sigma(E,\lambda)\cdot \pi_{\lambda *}\left((\times_{p\in P^*}\theta_p)^*(\times_{p\in P^*}\mathrm{vol}_{E_p})\right) \\
&\overset{\text{(vii)}}{=} \sum_{\lambda\in\mathrm{EssCond}(\Gamma)} \sigma(\Gamma,\lambda)\cdot \times_{p\in P^*}\left(\widehat{\mathrm{I}}(\Gamma(\lambda,p))\right) \\
&\overset{\text{(viii)}}{=} \sum_{\lambda\in\mathrm{Cond}(\Gamma)} (\times_{p\in P^*}\widehat{\mathrm{I}})(\Gamma(\lambda)) \\
&\overset{\text{(ix)}}{=} \left(\times_{p\in P^*}\widehat{\mathrm{I}}\right)(\widehat{\Psi}_\nu(\Gamma)).
\end{aligned}$$

The justifications are

(i): by definition of $\widehat{\mathrm{I}}$;
(ii): by pullback formula of the pushforward [**18**, Proposition 8.9];
(iii): by Propositions 5.6.2 and 5.6.6, and additivity of the pushforward [**18**, Proposition 8.11];
(iv): by Proposition 5.6.5 (ii) and naturality of the pushforward [**18**, Proposition 8.10];
(v): by commutativity of (9.28);
(vi): by Lemma 9.5.2;
(vii): by definition of $\sigma(\Gamma,\lambda)$ in (7.4), definition of π_λ in (5.30), and Lemma 9.5.3 (i);
(viii): by definition of $\Gamma(\lambda)$ in (7.5) and Lemma 9.5.3(ii);
(ix): by definition of $\widehat{\Psi}_\nu$ in (7.6).

This finishes the proof of Proposition 9.5.1, showing that $\widehat{\mathrm{I}}$ and I are almost morphisms of cooperads.

CHAPTER 10

Proofs of the formality theorems

In this chapter we prove all the formality theorems given in the Introduction. Here $\mathbb{K}$ is the field of real numbers $\mathbb{R}$.

For (non-relative) formality, the case of ambient dimension $N = 1$ is trivial because the little intervals operad is weakly equivalent to the associative operad which is clearly formal. Assume that $N \geq 2$. Let us show first that

$$\mathrm{I}\colon \mathcal{D}(A) \longrightarrow \Omega_{PA}(\mathrm{C}[A])$$

is a weak equivalence. It is a CDGA map by Corollary 9.3.3 and Proposition 9.4.1. The map induced in cohomology is surjective because, for a, b distinct in A, the single-chord diagrams $\Gamma\langle a, b\rangle$ defined in (8.2) are sent to $\theta^*_{ab}(\mathrm{vol})$ which correpond clearly to the generators g_{ab} of the cohomology algebra of the configuration space (see Chapter 8). Since by Theorem 8.1 $\mathrm{H}(\mathcal{D}(A)) \cong \mathrm{H}^*(\mathrm{C}[A]))$, we deduce that I is a quasi-isomorphism.

As reviewed in Chapter 4, by [**18**, Theorem 7.1] Ω_{PA} and $A_{PL}(u(-);\mathbb{R})$ are weakly equivalent symmetric monoidal contravariant functors where

$$u : \mathrm{CompactSemiAlg} \longrightarrow \mathrm{Top} \tag{10.1}$$

is the forgetful functor which is symmetric strongly monoidal. In view of Definition 3.1, all of this combined with Theorem 8.1 and Proposition 9.5.1 implies that, for $N \geq 3$, $\mathrm{H}(\mathrm{C}[\bullet];\mathbb{R})$ is a CDGA model for the operad $\mathrm{C}[\bullet]$, and hence the same is true for the little N-disks operad. This establishes Theorem 1.2, that is, the formality of the little balls operad over $\mathbb{R}$ in the sense of Definitions 3.1 and 3.2.

When $N = 2$, the above argument does not prove the formality because $\mathcal{D}_2$ is only a cooperad of $\mathbb{Z}$-graded CDGAs (see end of Remark 6.5.6) and is therefore not suitable for modeling rational (or real) homotopy theory. However, we do have a zig-zag of quasi-isomorphisms of $\mathbb{Z}$-graded CDGA (almost) cooperads between $\mathrm{H}^*(\mathrm{C}_2[\bullet];\mathbb{R})$ and $\Omega^*_{PA}(\mathrm{C}_2[\bullet])$. Moreover, if we replace the zeroth term of the little disks operad (corresponding to operations in arity 0) by the empty space and replace $\mathcal{D}(0)$ by 0, then it is possible to truncate the cooperad $\mathcal{D}$ by an acyclic operadic ideal to make it a connected CDGA and recover formality. However, we will not pursue this here.

We now deduce the stable formality of the operad, which is the formality in the category of operads of chain complexes. We assume $N \geq 2$. Recall from [**18**, Definition 3.1] the chain complex of semi-algebraic chains

$$\mathrm{C}_*\colon \mathrm{SemiAlg} \longrightarrow \mathrm{Ch}_{\mathbb{Z}},$$

which is monoidal. We define the $\mathbb{R}$-dual of a graded real vector space or of a graded $\mathbb{Z}$-module V as

$$V^\vee := \mathrm{Hom}(V, \mathbb{R}), \tag{10.2}$$

and denote the dual of a linear map $f\colon V \to W$ by $f^\vee\colon W^\vee \to V^\vee$. There is a natural pairing

$$\begin{aligned}(10.3)\qquad \langle -,-\rangle\colon \Omega_{PA}(X)\otimes \mathrm{C}(X) &\longrightarrow \mathbb{R}\\ \omega\otimes\gamma &\longmapsto \langle\omega,\gamma\rangle\end{aligned}$$

and, by [**18**, Proposition 7.3], the evaluation map

$$\begin{aligned}\mathrm{ev}\colon \mathrm{C}_*(X)\otimes\mathbb{R} &\xrightarrow{\simeq} (\Omega_{PA}(X))^\vee\\ \gamma &\longmapsto \langle -,\gamma\rangle\end{aligned}$$

is a monoidal symmetric weak equivalence when X is a compact semi-algebraic set.

Fix a weak ordered partition $\nu\colon A \to P$ and set $P^* = \{0\} \otimes P$, $A_p = \nu^{-1}(p)$, and $A_0 = P$ as in the setting 2.4.1. Consider the following diagram (in which we write $\otimes_{P^*}$ for $\otimes_{p\in P^*}$)

$$\begin{array}{ccccc}
\bigotimes_{P^*} \mathrm{C}_*(\mathrm{C}[A_p])\otimes\mathbb{R} & \xrightarrow[\simeq]{\otimes_{P^*}\mathrm{ev}} & \bigotimes_{P^*}(\Omega_{PA}(\mathrm{C}[A_p]))^\vee & \xrightarrow[\simeq]{\otimes_{P^*}(\mathrm{I}^\vee)} & \bigotimes_{P^*}(\mathcal{D}(A_p))^\vee\\
 & & \downarrow{\simeq} & & \downarrow{\simeq}\\
\times\ \Big\downarrow\ \simeq & & \Big(\bigotimes_{P^*}\Omega_{PA}(\mathrm{C}[A_p])\Big)^\vee & \xrightarrow[\simeq]{(\otimes_{P^*}\mathrm{I})^\vee} & \Big(\bigotimes_{P^*}\mathcal{D}(A_p)\Big)^\vee\\
 & & \simeq\ \uparrow\ (\times)^\vee & & \\
\mathrm{C}_*\Big(\prod_{p\in P^*}\mathrm{C}[A_p]\Big)\otimes\mathbb{R} & \xrightarrow[\simeq]{\mathrm{ev}} & \Big(\Omega_{PA}((\prod_{p\in P^*}\mathrm{C}[A_p])\Big)^\vee & & \Big\downarrow (\Psi_\nu)^\vee\\
\mathrm{C}_*(\Phi_\nu)\downarrow & & \downarrow(\Omega_{PA}(\Phi_\nu))^\vee & & \\
\mathrm{C}_*(\mathrm{C}[A])\otimes\mathbb{R} & \xrightarrow[\simeq]{\mathrm{ev}} & (\Omega_{PA}((\mathrm{C}[A]))^\vee & \xrightarrow[\simeq]{\mathrm{I}^\vee} & (\mathcal{D}(A))^\vee
\end{array}$$

This diagram is commutative by [**18**, Proposition 7.3] and Proposition 9.5.1.

Note that $\mathcal{D}^\vee$, as the dual of the cooperad of $\mathbb{Z}$-graded differential vector spaces $\mathcal{D}$, is an operad. The above diagram implies that the operad $\mathrm{C}_*(\mathrm{C}[\bullet])\otimes\mathbb{R}$ is weakly equivalent to $\mathcal{D}^\vee$. By Theorem 8.1, the latter is weakly equivalent to $H_*(\mathrm{C}[\bullet])\otimes\mathbb{R}$. By [**18**, Proposition 7.2], the symmetric monoidal functors of semi-algebraic chains C_* and of singular chains S_* are weakly equivalent. This proves Theorem 1.1, the stable formality of the little N-disks operad which says that the chains and the homology of the little balls operad are quasi-isomorphic.

We now arrive to the proof of the relative formality. Let $1 \le m < N$ be integers. Suppose given a linear isometry

$$\epsilon\colon \mathbb{R}^m \longrightarrow \mathbb{R}^N.$$

For an integer $d \ge 1$ and a finite set A, denote by $\mathrm{C}_d[A]$ the Fulton-MacPherson space of configurations in $\mathbb{R}^d$. Define the map

$$\mathrm{C}_\epsilon[A]\colon \mathrm{C}_m[A] \longrightarrow \mathrm{C}_N[A]$$

which sends a configuration in $\mathbb{R}^m$ to its image under ϵ in $\mathbb{R}^N$. Clearly this map induces a morphism of operads and is equivalent to the morphism induced by ϵ between the little balls operads.

Define the morphism between CDGAs of admissible diagrams in dimensions N and m

$$\mathcal{D}_\epsilon \colon \mathcal{D}_N(A) \longrightarrow \mathcal{D}_m(A)$$

by, for a diagram Γ in $\mathcal{D}_N(A)$,

$$\mathcal{D}_\epsilon(\Gamma) = \begin{cases} \mathbf{1}, & \text{if } \Gamma \text{ is the unit diagram;} \\ 0, & \text{otherwise.} \end{cases}$$

The case $m = 1$, however, is special. We set

$$\mathcal{D}_1(A) := \mathrm{H}^*(\mathrm{C}_1[A]) \cong \mathbb{R}[\mathrm{Perm}(A)],$$

hence $\mathcal{D}_1$ is the associative cooperad. The unit $\mathbf{1} \in \mathcal{D}_1(A)$ is the constant cohomology class $1 \in \mathrm{H}^0(\mathrm{C}[A])$, and $\mathcal{D}_\epsilon$ is defined in the same way as above.

LEMMA 10.1. *$\mathcal{D}_\epsilon$ is a morphism of CDGA cooperads and it is weakly equivalent to*

$$\mathrm{H}^*(\mathrm{C}_\epsilon[\bullet]) \colon \mathrm{H}^*(\mathrm{C}_N[\bullet]) \longrightarrow \mathrm{H}^*(\mathrm{C}_m[\bullet]).$$

PROOF. It is clear that $\mathcal{D}_\epsilon$ is a morphism of CDGA cooperads. Since $m < N$, $\mathrm{H}^*(\mathrm{C}_\epsilon; \mathbb{R})$ is the trivial map, that is, it is zero in positive degrees and maps the unit of the cohomology algebra to the unit. The same is true for $\mathcal{D}_\epsilon$. This, combined with Theorem 8.1 (which is tautological for $m = 1$) implies the result. □

We now want to prove that $\mathcal{D}_\epsilon$ is weakly equivalent to $\Omega_{PA}(\mathrm{C}_\epsilon)$. For this we will use the Kontsevich configuration space integral that we extend in ambient dimension $m = 1$ as the linear map

$$\mathrm{I} = \mathrm{I}_1 \colon \mathcal{D}_1(A) = \mathrm{H}^*(\mathrm{C}_1[A]) \longrightarrow \Omega^*_{PA}(\mathrm{C}_1[A])$$

that sends a (degree 0) cohomology class to the corresponding locally constant function on $\mathrm{C}_1[A]$. It is clearly a weak equivalence of "almost" cooperads as in Proposition 9.5.1.

We need the following

LEMMA 10.2. *Assume that $m \geq 1$ and $N \geq 2m+1$. Then the following diagram commutes:*

$$\begin{array}{ccc} \mathcal{D}_N(A) & \xrightarrow[\simeq]{\mathrm{I}_N} & \Omega_{PA}(\mathrm{C}_N[A]) \\ {\scriptstyle \mathcal{D}_\epsilon}\big\downarrow & & \big\downarrow{\scriptstyle \Omega_{PA}(\mathrm{C}_\epsilon)} \\ \mathcal{D}_m(A) & \xrightarrow[\simeq]{\mathrm{I}_m} & \Omega_{PA}(\mathrm{C}_m[A]). \end{array} \tag{10.4}$$

PROOF. The case $A = \emptyset$ is clear. Assume that $|A| \geq 1$. Let Γ be an admissible diagram in $\mathcal{D}_N(A)$. We have to show that

$$\Omega_{PA}(\mathrm{C}_\epsilon)\left(\mathrm{I}_N(\Gamma)\right) = \mathrm{I}_m\left(\mathcal{D}_\epsilon(\Gamma)\right), \tag{10.5}$$

where the right hand side is zero except when Γ is the unit diagram. For the unit diagram this is clear, so assume that Γ is not a unit and let us show that the left hand side of (10.5) vanishes. Denote Γ's set of edges by E and by I its set of internal vertices. Suppose first that moreover each external vertex of Γ is an endpoint of some edge. Since internal vertices are at least trivalent, this implies that

$$|E| \geq \frac{1}{2}(|A| + 3 \cdot |I|).$$

Since $N \geq 2m+1 \geq 3$, we deduce that

$$\begin{aligned}\deg(\Gamma) &= (N-1)\cdot|E| - N\cdot|I| \\ &\geq \frac{N-1}{2}(|A|+3\cdot|I|) - N\cdot|I| \\ &= \frac{N-3}{2}\cdot|I| + \frac{N-1}{2}\cdot|A| \\ &\geq |A|\cdot m \\ &> \dim(\mathrm{C}_m[A]),\end{aligned}$$

and hence the left hand side of Equation (10.5) vanishes for degree reasons.

Consider now a general admissible non-unit diagram Γ on A and let $B \subset A$ be the set of external vertices that are the endpoints of some edge of Γ. We have an obvious associated map

$$\begin{aligned}\iota\colon \mathcal{D}_N(B) &\longrightarrow \mathcal{D}_N(A) \\ \Gamma' &\longmapsto \iota(\Gamma')\end{aligned}$$

defined by adding to a diagram Γ' in $\mathcal{D}_N(B)$ isolated external vertices labeled by $A \setminus B$. Thus Γ is the image under ι of some diagram $\Gamma' \in \mathcal{D}_N(B)$. The map ι can easily be described in terms of cooperadic operations analogously to the operadic description before Definition 5.3.1 of the canonical projection

$$\pi : \mathrm{C}[A] \longrightarrow \mathrm{C}[B].$$

The following diagram commutes

$$\begin{array}{ccc}\mathcal{D}_N(B) & \xrightarrow[\simeq]{\mathrm{I}_N} & \Omega_{PA}(\mathrm{C}_N[B]) \\ \Big\downarrow{\scriptstyle\iota} & & \Big\downarrow{\scriptstyle\Omega_{PA}(\pi)} \\ \mathcal{D}_N(A) & \xrightarrow[\simeq]{\mathrm{I}_N} & \Omega_{PA}(\mathrm{C}_N[A]).\end{array}$$

Since each external vertex of Γ' is the endpoint of an edge, we get by the discussion above that

$$\Omega_{PA}(\mathrm{C}_\epsilon)\mathrm{I}_N(\Gamma') = \mathrm{I}_m\mathcal{D}_\epsilon(\Gamma').$$

The commutativity of the last diagram and naturality imply then that Equation (10.5) holds for $\Gamma = \iota(\Gamma')$. This proves the lemma. □

Finally we prove the last statement of the Introduction:

PROOF OF THEOREM 1.4. The last two lemmas clearly imply formality of the morphism C_ϵ, and hence the same for the corresponding map between operads of little balls (when $m \neq 2$).

The stable formality of the morphism of operads $\mathrm{C}_\epsilon[\bullet]$ is deduced from the unstable formality above exactly as in the absolute case. □

Index of notation

For the convenience of the reader, we include a short index of the most important notation. Each entry is followed by a short description and a reference to where the notation is defined in the paper.

Names and latin letters

A, A_Γ .... set of external vertices in a configuration space or in a diagram; Chapter 5, Definition 6.1.1 (see also A_p, I, V)

A_p .... $A_p = \nu^{-1}(p)$; Setting 2.4.1

A_0 .... alternative notation for the codomain P of a weak partition $\nu\colon A \to P$; Setting 2.4.1

AdmCond(Γ) .... set of admissible condensations on a diagram Γ; Definition 7.3.3

A_{PL} .... Sullivan functor of polynomial forms; Chapter 3

$\mathcal{B}_N(n)$, $\mathcal{B}_N(\bullet)$, $\mathcal{B}$ .... little N-disk operad; Chapter 1

barycenter(x) .... barycenter of a configuration in $\mathbb{R}^N$; (5.3)

$\mathcal{BF}(V)$ .... indexing set of the boundary faces of $\mathrm{C}[V]$; (5.12)

$\mathcal{BF}(V, A)$ .... indexing set of the boundary faces of the fiberwise boundary $\mathrm{C}^\partial[V]$; (5.38)

$\mathrm{C}_*(X)$, $\mathrm{C}_k(X)$ .... semi-algebraic chains on a semi-algebraic set X; Chapter 4

$\mathrm{C}(A)$, $\mathrm{C}(n)$ .... space of normalized configurations, identified with $\mathrm{Inj}_0^1(A, \mathbb{R}^N)$; (5.2), (5.5)

$\mathrm{C}[A]$, $\mathrm{C}[V]$, $\mathrm{C}[\bullet]$ .... Fulton-MacPherson compactification of configuration spaces and corresponding operad; Chapter 5, Definition 5.1.1

$\mathrm{C}[V, \nu]$ .... set of ν-condensed configurations; (5.15)

$\mathrm{C}^\partial[V]$ .... fiberwise boundary of $\pi\colon \mathrm{C}[V] \to \mathrm{C}[A]$; (5.37)

CDGA, $\mathrm{CDGA}_\mathbb{K}$ .... category of ($\mathbb{N}$-graded) commutative differential graded algebras over the field $\mathbb{K}$; beginning of Chapter 3

$\mathrm{Ch}_\mathbb{K}$ .... category of chain complexes with coefficients in $\mathbb{K}$

$\mathrm{C}^{\mathrm{sing}}(V_1, V_2)$ .... singular configuration space; (5.13)

Cond(V, ν), Cond(V) .... set of condensations on V relative to a weak partition $\nu\colon A \to P$; Definition 5.6.1

Cond(Γ, ν), Cond(Γ) .... set of condensations on V_Γ relative to a weak partition $\nu\colon A \to P$; Definition 7.1.2

d .... differential of a diagram; (6.1)

$\mathcal{D}(n)$, $\mathcal{D}_N(n)$, $\mathcal{D}(A)$, $\mathcal{D}(\bullet)$ .... spaces and cooperad of admissible diagrams; Definition 6.5.4

$\widehat{\mathcal{D}}(n)$, $\widehat{\mathcal{D}}_N(n)$, $\widehat{\mathcal{D}}(A)$, $\widehat{\mathcal{D}}(\bullet)$ .... spaces and cooperad of diagrams; Definition 6.2.2

$\deg(\Gamma)$	degree of a diagram; Definition 6.2.3
E_Γ	ordered set of edges of a diagram Γ; Definition 6.1.1
$E_\Gamma^{\mathrm{contr}}$	set of contractible edges of a diagram; Definition 6.1.1
$\mathrm{EssCond}(V)$, $\mathrm{EssCond}(V,\nu)$	set of essential condensations; Definition 5.6.1
g_{ab}	standard generator of the cohomology algebra of $\mathrm{C}[A]$; (8.1).
$\mathrm{I}, \widehat{\mathrm{I}}$	Kontsevich configuration space integrals; Chapter 9, (9.6), (9.7), Corollary 9.3.3
$\bar{\mathrm{I}}$	quasi-isomorphism between $\mathcal{D}(A)$ and $\mathrm{H}^*(\mathrm{C}[A])$; Theorem 8.1, (8.3)
I, I_Γ	set of vertices on a configuration space or ordered sets of internal vertices of a diagram; beginning of Section 5.3, Definition 6.1.1 (see also A and V)
I_p	set of p-local internal vertices (or global if $p=0$), $I_p = I \cap \lambda^{-1}(p)$; (5.19), after (7.1) (see also A_p, V_p)
I_0	set of global internal vertices, $I_0 = \lambda^{-1}(0)$; (5.19), after (7.1) (see also A_0, V_0)
$\mathrm{Inj}(A, \mathbb{R}^N)$	space of injections of A into $\mathbb{R}^N$; (5.1)
$\mathrm{Inj}_0^1(A, \mathbb{R}^N)$	space of injections of A into $\mathbb{R}^N$ with barycenter at the origin and radius 1, identified with $\mathrm{C}(A)$; (5.5)
$\mathbb{K}$	ground unital ring (often $\mathbb{K} = \mathbb{R}$); Chapter 2
N	fixed positive integer giving the ambient dimension of the little disks operad or the configuration space; Chapter 2
$\mathcal{N}(A)$	ideal of non-admissible diagrams on A; Definition 6.5.1
$\mathcal{N}(\nu)$	ideal of non-admissible diagrams associated to a weak partition ν; (7.7)
P	codomain of a weak (ordered) partition; Definition 2.3.1, Setting 2.4.1
P^*	extended codomain $P^* := \{0\} \olessthan P$ of a partition; Setting 2.4.1
$\mathrm{Perm}(A)$	set of permutations of a set A; Chapter 2
pos, $\mathrm{pos}(x : L)$	position function; Section 2.2
$\mathrm{radius}(x)$	radius of a configuration; (5.4)
$s_\Gamma(e)$	source of an edge; Definition 6.1.1
S^{N-1}	unit sphere in $\mathbb{R}^N$; (5.6)
$S(I,\lambda)$, $S(E,\lambda)$	sets used to define signs $\sigma(I,\lambda)$, $\sigma(E,\lambda)$; (5.33), preceeding (7.2)
$\mathrm{S}_*(-;\mathbb{K})$; S_*	functor of singular chains with coefficients in $\mathbb{K}$
SemiAlg, CompactSemiAlg	category of (compact) semi-algebraic sets; Definition 4.1
$t_\Gamma(e)$	target of an edge; Definition 6.1.1
Top	category of topological spaces
u	forgetful functor CompactSemiAlg $\to$ Top; Chapter 4, (10.1)
V, V_Γ	set of vertices of a configuration or a diagram, $V = A \amalg I$, $V_\Gamma = A_\Gamma \amalg I_\Gamma$; Definition 6.1.1
V_p	set of p-local internal vertices (or global if $p=0$), $V_p = A_p \cup I_p$; (5.19), after (7.1) (see also A_p, I_p)
V_0	set of global internal vertices, $V_0 = \lambda^{-1}(0) \cup P$; (5.19), after (7.1) (see also A_0, I_0)
vol	symmetric volume form on the unit sphere; (9.1)

vol_E .... volume form on a product of a family of spheres indexed by E; (9.2)

Greek letters

Γ .... a diagram, or an isomorphism class of diagram, or an equivalence class of diagram; Definition 6.1.1

$\Gamma\langle a,b\rangle$ .... a diagram consisting of a single chord joining the external vertices a and b; (8.2)

$\Gamma(\lambda)$, $\Gamma(\lambda,p)$ .... used to define the cooperadic structure on diagrams; (7.1), (7.5)

$\delta_{a,b,c}$ .... relative distance between three points of a configuration; (5.7)

$\epsilon(\Gamma,e)$ .... sign associated to the contraction of an edge e in a diagram; preceeding (6.1)

$\theta_{a,b}$, θ_{ab} .... map $\mathrm{C}[A]\to S^{N-1}$ giving the direction between two points of a configuration; (5.6)

θ_e .... θ-function associated to an edge e, $\theta_e=\theta_{s_\Gamma(e),t_\Gamma(e)}$; before (9.3)

θ_Γ .... product of maps θ_e indexed by the edges e of Γ; (9.3)

κ .... Kunneth quasi-isomorphism; (3.2), (3.4)

λ .... condensation $\lambda\colon V\to P^*$; Definition 5.6.1

λ_E .... extension of the condensation λ to edges; Definition 7.1.2, following (7.1)

$\widehat{\lambda}$ .... condensation associated to λ; (5.22), below (7.1)

ν .... weak partition $A\to P$; Definition 2.3.1, Setting 2.4.1

π .... canonical projection $\mathrm{C}[V]\to\mathrm{C}[A]$ for $A\subset V$; (5.10), Definition 5.3.1

π_* .... pushforward or integration along the fiber $\Omega^{k+*}_{\min}(E)\to\Omega^*_{PA}(B)$; (4.9)

π_Γ .... canonical projection $\mathrm{C}[V_\Gamma]\to\mathrm{C}[A_\Gamma]$; (9.4).

$(\pi_\Gamma)_*$ .... pushforward along the canonical projection π_Γ; (9.5)

$\pi^\partial\colon E^\partial\to B$ .... fiberwise boundary of an SA bundle π; (4.8)

$\sigma(I,\lambda)$, $\sigma(E,\lambda)$, $\sigma(\Gamma,\lambda)$ .... signs; (5.32), (7.2), (7.3), (7.4)

Φ_ν .... operadic structure map in $\mathrm{C}[\bullet]$ associated to a weak partition ν; (5.9)

$\Phi^V_W=\Phi_W$ .... operadic structure map corresponding to a circle operation $\mathrm{C}[V/W]\times\mathrm{C}[W]\to\mathrm{C}[V]$ for $W\subset V$; (5.11)

Ψ_ν .... cooperadic structure map on $\mathcal{D}(\bullet)$ associated to a weak partition ν; Section 7.1, Proposition 7.1.5

$\widehat{\Psi}_\nu$ .... cooperadic structure map on $\widehat{\mathcal{D}}(\bullet)$ associated to a weak partition ν; Section 7.1, (7.6)

$\Omega_{\min}$ .... functor of minimal forms on semi-algebraic sets; (4.3), (4.9)

Ω_{PA} .... functor of PA forms on semi-algebraic sets; (4.6), Theorem 4.2, (4.9)

Other symbols

$x(a)\simeq x(b)\,\mathrm{rel}\,x(c)$ .... proximity relation in $\mathrm{C}[A]$; (5.8)

$\Gamma \simeq \pm\Gamma'$	equivalence relation of diagrams; Definition 6.2.2
$[\![M]\!]$, $g_*([\![M]\!])$, $[\![\pi^{-1}(b)]\!]$	semi-algebraic chain represented by a compact semi-algebraic manifold M, its image by a semi-algebraic map g, or semi-algebraic chain represented by a fiber of an oriented SA bundle; (4.1), (4.7)
$\underline{n}$	set $\{1, \ldots, n\}$; Section 2.1
$f\|A$	restriction of a function to a subdomain; Section 2.1
$L_1 \oslash L_2$, $\oslash_{p\in P} L_p$	ordered sum; Section 2.2
$\langle -, - \rangle$	evaluation of a form; (4.4), (10.3)
Y^X	set of functions from X to Y; Section 2.1
$\|A\|$	cardinality of a set A; Section 2.1
$\overline{E}$	closure of a subset E in a topological space
$\amalg$	disjoint union of sets
V/W	quotient of a set V by a subset $W \subset V$; Section 5.4
Γ/e	contraction of an edge in a diagram; Definition 6.4.1
λ/e	condensation induced on a contracted diagram; (7.12).
$V^\vee$, $f^\vee$	linear dual of a vector space or of a linear map; (10.2)
$\\|x\\|$	Euclidean norm of $x \in \mathbb{R}^N$
$\wedge Z$	free commutative graded algebra generated by the graded vector space Z; after (8.1)

Bibliography

[1] V. I. Arnol′d, *The cohomology ring of the group of dyed braids*, Mat. Zametki **5** (1969), 227–231 (Russian). MR0242196 (39 #3529)

[2] Greg Arone, Pascal Lambrechts, Victor Turchin, and Ismar Volić, *Coformality and rational homotopy groups of spaces of long knots*, Math. Res. Lett. **15** (2008), no. 1, 1–14. MR2367169 (2008m:57059)

[3] Gregory Arone, Pascal Lambrechts, and Ismar Volić, *Calculus of functors, operad formality, and rational homology of embedding spaces*, Acta Math. **199** (2007), no. 2, 153–198, DOI 10.1007/s11511-007-0019-7. MR2358051 (2008k:18007)

[4] J. M. Boardman and R. M. Vogt, *Homotopy invariant algebraic structures on topological spaces*, Lecture Notes in Mathematics, Vol. 347, Springer-Verlag, Berlin, 1973. MR0420609 (54 #8623a)

[5] J. Bochnak, M. Coste, and M.-F. Roy, *Géométrie algébrique réelle*, Ergebnisse der Mathematik und ihrer Grenzgebiete (3) [Results in Mathematics and Related Areas (3)], vol. 12, Springer-Verlag, Berlin, 1987 (French). MR949442 (90b:14030)

[6] Raoul Bott and Clifford Taubes, *On the self-linking of knots*, J. Math. Phys. **35** (1994), no. 10, 5247–5287, DOI 10.1063/1.530750. Topology and physics. MR1295465 (95g:57008)

[7] A. K. Bousfield and V. K. A. M. Gugenheim, *On* PL *de Rham theory and rational homotopy type*, Mem. Amer. Math. Soc. **8** (1976), no. 179, ix+94. MR0425956 (54 #13906)

[8] Alberto S. Cattaneo, Paolo Cotta-Ramusino, and Riccardo Longoni, *Configuration spaces and Vassiliev classes in any dimension*, Algebr. Geom. Topol. **2** (2002), 949–1000 (electronic), DOI 10.2140/agt.2002.2.949. MR1936977 (2004a:57014)

[9] Fred Cohen, *Cohomology of braid spaces*, Bull. Amer. Math. Soc. **79** (1973), 763–766. MR0321074 (47 #9607)

[10] Edward Fadell and Lee Neuwirth, *Configuration spaces*, Math. Scand. **10** (1962), 111-118. MR0141126 (25 #4537)

[11] Yves Félix, Stephen Halperin, and Jean-Claude Thomas, *Rational homotopy theory*, Graduate Texts in Mathematics, vol. 205, Springer-Verlag, New York, 2001. MR1802847 (2002d:55014)

[12] William Fulton and Robert MacPherson, *A compactification of configuration spaces*, Ann. of Math. (2) **139** (1994), no. 1, 183–225, DOI 10.2307/2946631. MR1259368 (95j:14002)

[13] Giovanni Gaiffi, *Models for real subspace arrangements and stratified manifolds*, Int. Math. Res. Not. **12** (2003), 627–656, DOI 10.1155/S1073792803209077. MR1951400 (2004d:52021)

[14] E. Getzler and J. D. S Jones. Operads, homotopy algebra, and iterated integrals for double loop spaces. Preprint arXiv:hep-th/9403055v1.

[15] Victor Ginzburg and Mikhail Kapranov, *Koszul duality for operads*, Duke Math. J. **76** (1994), no. 1, 203–272, DOI 10.1215/S0012-7094-94-07608-4. MR1301191 (96a:18004)

[16] Thomas G. Goodwillie and Michael Weiss, *Embeddings from the point of view of immersion theory. II*, Geom. Topol. **3** (1999), 103–118 (electronic), DOI 10.2140/gt.1999.3.103. MR1694808 (2000c:57055b)

[17] F. Guillén Santos, V. Navarro, P. Pascual, and A. Roig, *Moduli spaces and formal operads*, Duke Math. J. **129** (2005), no. 2, 291–335, DOI 10.1215/S0012-7094-05-12924-6. MR2165544 (2006e:14033)

[18] Robert Hardt, Pascal Lambrechts, Victor Turchin, and Ismar Volić, *Real homotopy theory of semi-algebraic sets*, Algebr. Geom. Topol. **11** (2011), no. 5, 2477–2545, DOI 10.2140/agt.2011.11.2477. MR2836293 (2012i:55011)

[19] Maxim Kontsevich, *Feynman diagrams and low-dimensional topology*, First European Congress of Mathematics, Vol. II (Paris, 1992), Progr. Math., vol. 120, Birkhäuser, Basel, 1994, pp. 97–121. MR1341841 (96h:57027)

[20] Maxim Kontsevich, *Operads and motives in deformation quantization*, Lett. Math. Phys. **48** (1999), no. 1, 35–72, DOI 10.1023/A:1007555725247. Moshé Flato (1937–1998). MR1718044 (2000j:53119)

[21] Maxim Kontsevich, *Deformation quantization of Poisson manifolds*, Lett. Math. Phys. **66** (2003), no. 3, 157–216, DOI 10.1023/B:MATH.0000027508.00421.bf. MR2062626 (2005i:53122)

[22] Maxim Kontsevich and Yan Soibelman, *Deformations of algebras over operads and the Deligne conjecture*, Conférence Moshé Flato 1999, Vol. I (Dijon), Math. Phys. Stud., vol. 21, Kluwer Acad. Publ., Dordrecht, 2000, pp. 255–307. MR1805894 (2002e:18012)

[23] Pascal Lambrechts, Victor Turchin, and Ismar Volić, *The rational homology of spaces of long knots in codimension* >2, Geom. Topol. **14** (2010), no. 4, 2151–2187, DOI 10.2140/gt.2010.14.2151. MR2740644 (2011m:57032)

[24] Joseph Neisendorfer and Timothy Miller, *Formal and coformal spaces*, Illinois J. Math. **22** (1978), no. 4, 565–580. MR0500938 (58 #18429)

[25] Paolo Salvatore, *Configuration spaces with summable labels*, Cohomological methods in homotopy theory (Bellaterra, 1998), Progr. Math., vol. 196, Birkhäuser, Basel, 2001, pp. 375–395. MR1851264 (2002f:55039)

[26] Pavol Ševera and Thomas Willwacher, *Equivalence of formalities of the little discs operad*, Duke Math. J. **160** (2011), no. 1, 175–206, DOI 10.1215/00127094-1443502. MR2838354 (2012j:18014)

[27] Dev P. Sinha, *Manifold-theoretic compactifications of configuration spaces*, Selecta Math. (N.S.) **10** (2004), no. 3, 391–428, DOI 10.1007/s00029-004-0381-7. MR2099074 (2005h:55015)

[28] Dev P. Sinha, *Operads and knot spaces*, J. Amer. Math. Soc. **19** (2006), no. 2, 461–486 (electronic), DOI 10.1090/S0894-0347-05-00510-2. MR2188133 (2006k:57070)

[29] Jim Stasheff, *What is ... an operad?*, Notices Amer. Math. Soc. **51** (2004), no. 6, 630–631. MR2064150

[30] Dennis Sullivan, *Infinitesimal computations in topology*, Inst. Hautes Études Sci. Publ. Math. **47** (1977), 269–331 (1978). MR0646078 (58 #31119)

[31] Dmitry E. Tamarkin, *Formality of chain operad of little discs*, Lett. Math. Phys. **66** (2003), no. 1-2, 65–72, DOI 10.1023/B:MATH.0000017651.12703.a1. MR2064592 (2005j:18010)

[32] Michael Weiss, *Embeddings from the point of view of immersion theory. I*, Geom. Topol. **3** (1999), 67–101 (electronic), DOI 10.2140/gt.1999.3.67. MR1694812 (2000c:57055a)

Editorial Information

To be published in the *Memoirs*, a paper must be correct, new, nontrivial, and significant. Further, it must be well written and of interest to a substantial number of mathematicians. Piecemeal results, such as an inconclusive step toward an unproved major theorem or a minor variation on a known result, are in general not acceptable for publication.

Papers appearing in *Memoirs* are generally at least 80 and not more than 200 published pages in length. Papers less than 80 or more than 200 published pages require the approval of the Managing Editor of the Transactions/Memoirs Editorial Board. Published pages are the same size as those generated in the style files provided for $\mathcal{A}\mathcal{M}\mathcal{S}$-LaTeX or $\mathcal{A}\mathcal{M}\mathcal{S}$-TeX.

Information on the backlog for this journal can be found on the AMS website starting from `http://www.ams.org/memo`.

A Consent to Publish is required before we can begin processing your paper. After a paper is accepted for publication, the Providence office will send a Consent to Publish and Copyright Agreement to all authors of the paper. By submitting a paper to the *Memoirs*, authors certify that the results have not been submitted to nor are they under consideration for publication by another journal, conference proceedings, or similar publication.

Information for Authors

Memoirs is an author-prepared publication. Once formatted for print and on-line publication, articles will be published as is with the addition of AMS-prepared frontmatter and backmatter. Articles are not copyedited; however, confirmation copy will be sent to the authors.

Initial submission. The AMS uses Centralized Manuscript Processing for initial submissions. Authors should submit a PDF file using the Initial Manuscript Submission form found at `www.ams.org/submission/memo`, or send one copy of the manuscript to the following address: Centralized Manuscript Processing, MEMOIRS OF THE AMS, 201 Charles Street, Providence, RI 02904-2294 USA. If a paper copy is being forwarded to the AMS, indicate that it is for *Memoirs* and include the name of the corresponding author, contact information such as email address or mailing address, and the name of an appropriate Editor to review the paper (see the list of Editors below).

The paper must contain a *descriptive title* and an *abstract* that summarizes the article in language suitable for workers in the general field (algebra, analysis, etc.). The *descriptive title* should be short, but informative; useless or vague phrases such as "some remarks about" or "concerning" should be avoided. The *abstract* should be at least one complete sentence, and at most 300 words. Included with the footnotes to the paper should be the 2010 *Mathematics Subject Classification* representing the primary and secondary subjects of the article. The classifications are accessible from `www.ams.org/msc/`. The Mathematics Subject Classification footnote may be followed by a list of *key words and phrases* describing the subject matter of the article and taken from it. Journal abbreviations used in bibliographies are listed in the latest *Mathematical Reviews* annual index. The series abbreviations are also accessible from `www.ams.org/msnhtml/serials.pdf`. To help in preparing and verifying references, the AMS offers MR Lookup, a Reference Tool for Linking, at `www.ams.org/mrlookup/`.

Electronically prepared manuscripts. The AMS encourages electronically prepared manuscripts, with a strong preference for $\mathcal{A}\mathcal{M}\mathcal{S}$-LaTeX. To this end, the Society has prepared $\mathcal{A}\mathcal{M}\mathcal{S}$-LaTeX author packages for each AMS publication. Author packages include instructions for preparing electronic manuscripts, samples, and a style file that generates the particular design specifications of that publication series. Though $\mathcal{A}\mathcal{M}\mathcal{S}$-LaTeX is the highly preferred format of TeX, author packages are also available in $\mathcal{A}\mathcal{M}\mathcal{S}$-TeX.

Authors may retrieve an author package for *Memoirs of the AMS* from `www.ams.org/journals/memo/memoauthorpac.html` or via FTP to `ftp.ams.org` (login as `anonymous`, enter your complete email address as password, and type `cd pub/author-info`). The

AMS Author Handbook and the *Instruction Manual* are available in PDF format from the author package link. The author package can also be obtained free of charge by sending email to tech-support@ams.org or from the Publication Division, American Mathematical Society, 201 Charles St., Providence, RI 02904-2294, USA. When requesting an author package, please specify AMS-LaTeX or AMS-TeX and the publication in which your paper will appear. Please be sure to include your complete mailing address.

After acceptance. The source files for the final version of the electronic manuscript should be sent to the Providence office immediately after the paper has been accepted for publication. The author should also submit a PDF of the final version of the paper to the editor, who will forward a copy to the Providence office.

Accepted electronically prepared files can be submitted via the web at www.ams.org/submit-book-journal/, sent via FTP, or sent on CD to the Electronic Prepress Department, American Mathematical Society, 201 Charles Street, Providence, RI 02904-2294 USA. TeX source files and graphic files can be transferred over the Internet by FTP to the Internet node ftp.ams.org (130.44.1.100). When sending a manuscript electronically via CD, please be sure to include a message indicating that the paper is for the *Memoirs*.

Electronic graphics. Comprehensive instructions on preparing graphics are available at www.ams.org/authors/journals.html. A few of the major requirements are given here.

Submit files for graphics as EPS (Encapsulated PostScript) files. This includes graphics originated via a graphics application as well as scanned photographs or other computer-generated images. If this is not possible, TIFF files are acceptable as long as they can be opened in Adobe Photoshop or Illustrator.

Authors using graphics packages for the creation of electronic art should also avoid the use of any lines thinner than 0.5 points in width. Many graphics packages allow the user to specify a "hairline" for a very thin line. Hairlines often look acceptable when proofed on a typical laser printer. However, when produced on a high-resolution laser imagesetter, hairlines become nearly invisible and will be lost entirely in the final printing process.

Screens should be set to values between 15% and 85%. Screens which fall outside of this range are too light or too dark to print correctly. Variations of screens within a graphic should be no less than 10%.

Inquiries. Any inquiries concerning a paper that has been accepted for publication should be sent to memo-query@ams.org or directly to the Electronic Prepress Department, American Mathematical Society, 201 Charles St., Providence, RI 02904-2294 USA.

Editors

This journal is designed particularly for long research papers, normally at least 80 pages in length, and groups of cognate papers in pure and applied mathematics. Papers intended for publication in the *Memoirs* should be addressed to one of the following editors. The AMS uses Centralized Manuscript Processing for initial submissions to AMS journals. Authors should follow instructions listed on the Initial Submission page found at www.ams.org/memo/memosubmit.html.

Algebra, to ALEXANDER KLESHCHEV, Department of Mathematics, University of Oregon, Eugene, OR 97403-1222; e-mail: klesh@uoregon.edu

Algebraic geometry, to DAN ABRAMOVICH, Department of Mathematics, Brown University, Box 1917, Providence, RI 02912; e-mail: amsedit@math.brown.edu

Algebraic topology, to SOREN GALATIUS, Department of Mathematics, Stanford University, Stanford, CA 94305 USA; e-mail: transactions@lists.stanford.edu

Arithmetic geometry, to TED CHINBURG, Department of Mathematics, University of Pennsylvania, Philadelphia, PA 19104-6395; e-mail: math-tams@math.upenn.edu

Automorphic forms, representation theory and combinatorics, to DANIEL BUMP, Department of Mathematics, Stanford University, Building 380, Sloan Hall, Stanford, California 94305; e-mail: bump@math.stanford.edu

Combinatorics, to JOHN R. STEMBRIDGE, Department of Mathematics, University of Michigan, Ann Arbor, Michigan 48109-1109; e-mail: JRS@umich.edu

Commutative and homological algebra, to LUCHEZAR L. AVRAMOV, Department of Mathematics, University of Nebraska, Lincoln, NE 68588-0130; e-mail: avramov@math.unl.edu

Differential geometry and global analysis, to CHRIS WOODWARD, Department of Mathematics, Rutgers University, 110 Frelinghuysen Road, Piscataway, NJ 08854; e-mail: ctw@math.rutgers.edu

Dynamical systems and ergodic theory and complex analysis, to YUNPING JIANG, Department of Mathematics, CUNY Queens College and Graduate Center, 65-30 Kissena Blvd., Flushing, NY 11367; e-mail: Yunping.Jiang@qc.cuny.edu

Functional analysis and operator algebras, to NATHANIEL BROWN, Department of Mathematics, 320 McAllister Building, Penn State University, University Park, PA 16802; e-mail: nbrown@math.psu.edu

Geometric analysis, to WILLIAM P. MINICOZZI II, Department of Mathematics, Johns Hopkins University, 3400 N. Charles St., Baltimore, MD 21218; e-mail: trans@math.jhu.edu

Geometric topology, to MARK FEIGHN, Math Department, Rutgers University, Newark, NJ 07102; e-mail: feighn@andromeda.rutgers.edu

Harmonic analysis, complex analysis, to MALABIKA PRAMANIK, Department of Mathematics, 1984 Mathematics Road, University of British Columbia, Vancouver, BC, Canada V6T 1Z2; e-mail: malabika@math.ubc.ca

Harmonic analysis, representation theory, and Lie theory, to E. P. VAN DEN BAN, Department of Mathematics, Utrecht University, P.O. Box 80 010, 3508 TA Utrecht, The Netherlands; e-mail: E.P.vandenBan@uu.nl

Logic, to ANTONIO MONTALBAN, Department of Mathematics, The University of California, Berkeley, Evans Hall #3840, Berkeley, California, CA 94720; e-mail: antonio@math.berkeley.edu

Number theory, to SHANKAR SEN, Department of Mathematics, 505 Malott Hall, Cornell University, Ithaca, NY 14853; e-mail: ss70@cornell.edu

Partial differential equations, to GUSTAVO PONCE, Department of Mathematics, South Hall, Room 6607, University of California, Santa Barbara, CA 93106; e-mail: ponce@math.ucsb.edu

Partial differential equations and functional analysis, to ALEXANDER KISELEV, Department of Mathematics, University of Wisconsin-Madison, 480 Lincoln Dr., Madison, WI 53706; e-mail: kisilev@math.wisc.edu

Probability and statistics, to PATRICK FITZSIMMONS, Department of Mathematics, University of California, San Diego, 9500 Gilman Drive, La Jolla, CA 92093-0112; e-mail: pfitzsim@math.ucsd.edu

Real analysis and partial differential equations, to WILHELM SCHLAG, Department of Mathematics, The University of Chicago, 5734 South University Avenue, Chicago, IL 60615; e-mail: schlag@math.uchicago.edu

All other communications to the editors, should be addressed to the Managing Editor, ALEJANDRO ADEM, Department of Mathematics, The University of British Columbia, Room 121, 1984 Mathematics Road, Vancouver, B.C., Canada V6T 1Z2; e-mail: adem@math.ubc.ca

Selected Published Titles in This Series

1078 **Milen Yakimov,** On the Spectra of Quantum Groups, 2013

1077 **Christopher P. Bendel, Daniel K. Nakano, Brian J. Parshall, and Cornelius Pillen,** Cohomology for Quantum Groups via the Geometry of the Nullcone, 2013

1076 **Jaeyoung Byeon and Kazunaga Tanaka,** Semiclassical Standing Waves with Clustering Peaks for Nonlinear Schrödinger Equations, 2013

1075 **Deguang Han, David R. Larson, Bei Liu, and Rui Liu,** Operator-Valued Measures, Dilations, and the Theory of Frames, 2013

1074 **David Dos Santos Ferreira and Wolfgang Staubach,** Global and Local Regularity of Fourier Integral Operators on Weighted and Unweighted Spaces, 2013

1073 **Hajime Koba,** Nonlinear Stability of Ekman Boundary Layers in Rotating Stratified Fluids, 2014

1072 **Victor Reiner, Franco Saliola, and Volkmar Welker,** Spectra of Symmetrized Shuffling Operators, 2014

1071 **Florin Diacu,** Relative Equilibria in the 3-Dimensional Curved n-Body Problem, 2014

1070 **Alejandro D. de Acosta and Peter Ney,** Large Deviations for Additive Functionals of Markov Chains, 2014

1069 **Ioan Bejenaru and Daniel Tataru,** Near Soliton Evolution for Equivariant Schrödinger Maps in Two Spatial Dimensions, 2014

1068 **Florica C. Cîrstea,** A Complete Classification of the Isolated Singularities for Nonlinear Elliptic Equations with Inverse Square Potentials, 2014

1067 **A. González-Enríquez, A. Haro, and R. de la Llave,** Singularity Theory for Non-Twist KAM Tori, 2014

1066 **José Ángel Peláez and Jouni Rättyä,** Weighted Bergman Spaces Induced by Rapidly Increasing Weights, 2014

1065 **Emmanuel Schertzer, Rongfeng Sun, and Jan M. Swart,** Stochastic Flows in the Brownian Web and Net, 2014

1064 **J. L. Flores, J. Herrera, and M. Sánchez,** Gromov, Cauchy and Causal Boundaries for Riemannian, Finslerian and Lorentzian Manifolds, 2013

1063 **Philippe Gille and Arturo Pianzola,** Torsors, Reductive Group Schemes and Extended Affine Lie Algebras, 2013

1062 **H. Inci, T. Kappeler, and P. Topalov,** On the Regularity of the Composition of Diffeomorphisms, 2013

1061 **Rebecca Waldecker,** Isolated Involutions in Finite Groups, 2013

1060 **Josef Bemelmans, Giovanni P. Galdi, and Mads Kyed,** On the Steady Motion of a Coupled System Solid-Liquid, 2013

1059 **Robert J. Buckingham and Peter D. Miller,** The Sine-Gordon Equation in the Semiclassical Limit: Dynamics of Fluxon Condensates, 2013

1058 **Matthias Aschenbrenner and Stefan Friedl,** 3-Manifold Groups Are Virtually Residually p, 2013

1057 **Masaaki Furusawa, Kimball Martin, and Joseph A. Shalika,** On Central Critical Values of the Degree Four L-Functions for GSp(4): The Fundamental Lemma. III, 2013

1056 **Bruno Bianchini, Luciano Mari, and Marco Rigoli,** On Some Aspects of Oscillation Theory and Geometry, 2013

1055 **A. Knightly and C. Li,** Kuznetsov's Trace Formula and the Hecke Eigenvalues of Maass Forms, 2013

1054 **Kening Lu, Qiudong Wang, and Lai-Sang Young,** Strange Attractors for Periodically Forced Parabolic Equations, 2013

For a complete list of titles in this series, visit the
AMS Bookstore at **www.ams.org/bookstore/memoseries/**.